South Asia
in the Globalised World

Edited by

C. Vinodan

Assistant Professor and Chair,
Centre for Strategic and Security Studies,
School of International Relations and Politics,
Mahatma Gandhi University, Kottayam, Kerala.

New Century Publications
New Delhi, India

NEW CENTURY PUBLICATIONS
4800/24, Bharat Ram Road,
Ansari Road, Daryaganj,
New Delhi - 110 002 (India)

Tel.: 011-2324 7798, 4358 7398, 6539 6605
Fax: 011-4101 7798
E-mail: indiatax@vsnl.com • info@newcenturypublications.com
www.newcenturypublications.com

Editorial office:
LG-7, Aakarshan Bhawan,
4754-57/23, Ansari Road, Daryaganj,
New Delhi - 110 002

Tel.: 011-4356 0919

First Published: **2014**

ISBN: **978-81-7708-347-7**

Published by New Century Publications and printed at Salasar Imaging Systems, New Delhi.

Designs: Patch Creative Unit, New Delhi.

PRINTED IN INDIA

About the Book

The world is experiencing strategic transition from the short-lived Post-Cold War unipolar system to a new and complex configuration of international relations. In the fast-changing global scenario, the importance of South Asia has come to be felt more than ever before, and across the world. South Asia, which consists of 8 states and more than one-fifth of humanity, has emerged as a vital region to the emerging global order due to its geo-strategic significance, growing economic dynamism, fragile state structures and crisis in governance.

This book contains 16 research articles which provide deep insights into the complex and multi-faceted nature of South Asia in the globalised world. These have been authored by scholars representing the disciplines of international relations, comparative politics, economics, security studies, and political economy. To understand the new realities of South Asia, the book would prove useful for researchers, policy makers, strategic analysts and others interested in world politics and international relations.

Editor's Profile

Dr. C. Vinodan is currently Assistant Professor and Chair, Centre for Strategic and Security Studies, School of International Relations and Politics, Mahatma Gandhi University, Kottayam, Kerala. He worked previously at the Post-graduate and Research Department of Political Science, Sree Narayana College, Kollam and Department of Political Science, University of Kerala. He obtained his M.A., M.Phil. and Ph.D. degrees in Political Science from the University of Kerala.

Dr. Vinodan was awarded Visiting Scholar Fellowship by the Claremont Graduate University, California, USA in 2003. He was also the recipient of Fulbright American Studies Institute (FASI) Fellowship in 2000 and subsequently completed a course on Foreign Policy at the University of Delaware, Philadelphia in 2000. He is currently member of the editorial board of the *Journal of Human Security*, *Journal of Politics and International Relations* and *South Asian Journal of Diplomacy*.

Contents

Contributors

Sabina W. Lautensach	Director, Human Security Institute, Canada.
Alexander K. Lautensach	Assistant Professor, School of Education, University of Northern British Columbia, Canada.
Mohanan B. Pillai	Professor and Director, UNESCO Madanjeet Singh Institute of South Asia Regional Co-operation (UMISARC), Pondicherry University, Puducherry.
P.M. Joshy	Guest Faculty, School of International Relations and Politics, Mahatma Gandhi University, Kottayam.
Dharmesh Pushpa Dhawankar	Senior Faculty, Department of Mass Communication, RTM, Nagpur University, Nagpur.
S.S. Kalamkar	Faculty Member, Gokhale Institute of Politics and Economics, Pune.
Girish Kumar R.	Centre for Trade, Development and Governance, School of International Relations and Politics, Mahatma Gandhi University, India.
Harsh V. Pant	Department of Defence Studies, King's College, London.
Shaijumon C.S.	Reader in Economics, Department of Humanities, Indian Institute of Space Science and Technology (IIST), Government of India.
M.J. Vinod	Professor, Department of Political Science, Bangalore University, Bangalore.
V. Mathew Kurian	Visiting Professor, School of International Relations and Politics, Mahatma Gandhi University, Kottayam, Kerala.

A.D. Manikandan Research Associate, Centre for Wage Employment and Poverty Alleviation, National Institute of Rural Development, Hyderabad.

G. Gopakumar Formerly Professor and Head, Department of Political Science, University of Kerala, Thiruvananthapuram.

Rahul Tripathi Assistant Professor, Department of Political Science, University of Goa, Goa.

Rajesh Basrur Senior Fellow, S. Rajaratnam School of International Studies, Nanyang Technological University, Singapore.

Anjali Sharma Associate Fellow, Observer Research Foundation, New Delhi.

K.M. Seethi Dean, Faculty of Social Sciences and Director of Research, Mahatma Gandhi University, Kottayam, Kerala.

C. Vinodan Assistant Professor and Chair, Centre for Strategic and Security Studies, School of International Relations and Politics, Mahatma Gandhi University, Kottayam, Kerala.

Shibu M. George Assistant Professor, Department of Political Science, Baselius College, Kottayam.

Preface

South Asia has emerged as a vital region in the emerging global order due to its geo-strategic significance, growing economic dynamism, fragile state structures and crisis in governance. For the first time in the last 350 years, the international system is undergoing a shift in its centre of gravity from the continents of Europe and North America to the rising Asia. Given the geographic proximity and economic complementarities between South Asia, on the one hand, and China, on the other, this region could become the greatest economic powerhouse in human history. The rising menace of transnational terrorism and nuclear proliferation further invited global attention to the region. For many years conflict and violence have been endemic to South Asia, a region comprising India, Pakistan, Sri Lanka, Bangladesh, Nepal, Bhutan, Afghanistan and the Maldives, leading to experts terming the region as 'the most dangerous place on Earth'.

Threats to security in South Asia stem from five main sources: (a) the unresolved nature of the Kashmir dispute and growing Indo-Pakistan hostility in the wake of the Mumbai terrorist attacks of November 2008; (b) strengthening of the Taliban insurgency in Afghanistan, with serious spill-over effects in Pakistan and to a lesser extent in India; (c) internal divisions and schisms in South Asian states leading to various types of violent intra-state 'identity conflicts'; (d) enormous spread of transnational terror networks in the region; and (e) failure of the South Asian Association for Regional Co-operation (SAARC) to promote regional integration, co-operation and peace.

South Asia is also a dynamic and complex region that presents the world with significant opportunities and challenges to global security, stability and human well-being. It is one of the fastest growing regions in the world with an average growth rate of above 6 percent over the past one decade and has demonstrated its potential for emerging as a significant economic entity in the world economic space. The region is home to India, the world's largest democracy and a rising power with one of fastest growing economies. It also includes governments struggling to maintain control of their countries, two nuclear powers that co-exist in an uneasy and often tense relationship, and conflicts with Al-Qaeda and Taliban forces.

These factors make the region a subject of considerable strategic importance to the major powers and the larger international community. South Asia covers 3.3 percent of Earth's terrestrial area, but supports more than one-fifth of the world's inhabitants. According to UN statistics, the region's population will rise by another 800 million people by 2050. It is the largest regional grouping in terms of human resources than that of EU and ASEAN. SAARC was formed in 1985 with the goal of raising the living standard of the region's poverty-stricken people. Unfortunately, it has not been able to leverage its full potential. Geography coupled with high levels of poverty and population density has rendered South Asia especially vulnerable to the impacts of climate change. This could compound development problems and further strain the resources needed to sustain growth, urbanisation, and industrialisation.

South Asia is also one of the least developed regions in the world. The rapid globalisation currently being experienced has prompted changes throughout the region. As a region with unique ethnic and religious background and historical processes, South Asia always has a significant place in world affairs. The countries of the region range in size from India, with a population of over 1 billion, to the Maldives with a population of just below 4,00,000. Globalisation has brought the region to crossroads.

Large sections of the South Asian people do not have access to education, health, nutrition, and such basic amenities like safe drinking water, sanitation and electricity. Hence, the region is facing acute developmental challenges. Economic analysis has made it clear that for realising any meaningful improvement in the welfare of the people of the South Asian region, the present growth rate of about 6 percent is not adequate. It is envisaged that a sustained growth rate of above 8 percent in GDP is necessary to facilitate the much needed structural transformation, so as to bring about the desired reduction in poverty rates and thereby improve the living conditions of the population of the region. The challenges of realizing and maintaining such high growth rates over a reasonable period of time are indeed complex and numerous. Among the most serious and pressing issue before the SAARC leaders is to balance the global climate challenges without compromising its developmental needs.

South Asia is also a crisis region because of the rivalry between Pakistan and India which began at the time of the founding of the state of Pakistan in 1947. The trauma of partition, communal violence and mass-scale population migration that accompanied the birth of India and Pakistan in 1947 was a portent of things to come. Relations between India and Pakistan have been marked by a series of crises which escalated into wars in 1947-48, 1965 and 1971. The 1971 War was not about Kashmir but resulted in the division of Pakistan, and this defeat intensified the sense of insecurity in Pakistan and the conflict between the two states.

There are emerging non-traditional security issues in South Asia that are of common concern to countries across the region. The 21st century is wrought with such "non-traditional" challenges as looming food and water resource crises, the often devastating environmental impacts of climate change, and the threat of pandemic diseases that cut across geographic boundaries. All of these, as witnessed in recent years, have an immediate human impact on South Asian countries, with implications for both future domestic and regional stability. Calamities in any one of these areas bear the very real potential to exacerbate conditions contributing to traditional security threats in the region.

The present volume consists of 16 research papers written by eminent scholars in the field from India and abroad. The book provides a multi-faceted analysis of the dynamics of South Asia.

Sabina W. Lautensach and Alexander K. Lautensach in their paper titled "Sustainable Development and Human Security: Global Context and South Asian Solutions" explain the relationship between sustainable development and human security and focus on South Asia especially on India's trajectory of sustainable development. They are sceptical about India's policy approach towards development which is still more or less not inclusive or unsustainable.

Mohanan B. Pillai and P.M. Joshy in "Social Capital and Region Building in South Asia" attempt to critically examine the notion of social capital in the context of the ascendant neo-liberal global regime and South Asian regional integration. The paper argues that a discussion about South Asian integration is incomplete without an understanding of the politics of identities.

Dharmesh Pushpa Dhawankar's paper "Ethnic Violence in South Asia: A Media Perspective" looks into various aspects of ethnic conflict around the globe especially in South Asian context. Most of the post-colonial societies are going through internal turmoil in the name of linguistics, national, religious, tribal, and racial divisions. The countries like India, Sri Lanka, Bangladesh, Nepal, Burma, and Pakistan are routinely seeing displacement of peoples due to ethnic conflict.

S.S. Kalamkar, in his paper "Food Security in South Asia", argues that food security is a growing international issue especially for developing countries in Asia, Africa and Latin America. The paper tries to study the extent of food insecurity in South Asian countries. He argues for the adoption of sustainable food security to suit the growing needs of the region.

Girish Kumar's paper "India's Innovation Systems: Changing Structure in the Post-TRIPS Regime" argues that

India is emerging as a major hub of R&D for multinational companies across the world due to the conducive policy adopted by the Government and the availability of large pool of intellectual resources.

Harsh V. Pant in his paper "The Trouble with SAARC" explains the unexpected challenges facing the SAARC countries from terrorism, extremism, poverty, climate change, narcotics and organized crime. According to him, progress in SAARC has stalled due to hesitancy borne from historical legacies, differences and disputes. The member states are unwilling to invest in SAARC and public perception of the organization is largely negative.

Shaijumon C.S. in his paper "Indo-US Trade Relations" argues that the relations between India and US entered into a new phase when both countries signed the agreement on 'Next Step in Strategic Partnership' in 2004. He argues that both countries should also address the direct investment and removal of certain barriers on outsourcing of services.

M.J. Vinod's paper "Obama Administration and US-Pakistan Relations" reviews the issues and trends in the relations between the two countries in the post-9/11 world. The paper argues that the strategic relationship between US and Pak entered into a new phase when Obama took over the administration. Currently, there exists distrust between both countries.

V. Mathew Kurian and A.D. Manikandan in their paper "India-Sri Lanka Free Trade Agreement in Spices" critically reviews the growth of regional trade agreements (RTAs) with specific reference to India-Sri Lanka Free Trade Agreement. The paper argues that the India-Sri Lanka Free Trade Agreement has brought benefits to both countries.

G. Gopa Kumar's paper "Future of Civil Society in Pakistan: The Post-Musharraf Era" tries to analyse the role played by the civil society in Pakistan which is going through internal turmoil. Relevance of civil society becomes crucial when it comes to the problem of governance in the postcolonial societies especially in the complex era of globalization.

Rahul Tripathi's paper "Political Economy of Pakistan: Security Implications for South Asia" argues that Pakistan currently faces one of the most challenging times in its history of 65 years. The paper also tries to explain the contours of the current crisis in Pakistan by looking at some important markers from a political economy perspective.

Rajesh Basrur's paper "Whither India's Bomb? Motivational Models Revisited" explains the trajectory of India's nuclear weapons programme. The paper analyses India's nuclear programme by using the three models identified by Scott Sagan—the security, the domestic politics, and the norms models—and argues that there is a high probability that the Indian nuclear weapons programme will continue to develop a wide range of operational capabilities at a pace dictated by its operational stakeholders, but restricted by the norms of the political leadership.

Anjali Sharma's paper "Redefining Regional Co-operation in South Asia" surveys the various dimensions of regional co-operation in South Asia in the new context. The paper also tries to understand the relationship between the region and its inhabitants and its impact on the success and failure of regional cooperation.

K.M. Seethi's paper "Negotiating for Security in South Asia" argues that security in the era of globalisation is a complex construct unfolding in diverse forms, embedded in internal, bilateral, regional and international settings. This is perhaps most evident in South Asia, which has been in its critical phase of transition during the past two decades.

C. Vinodan and Shibu M. George in their paper "Regional Integration and Energy Security in South Asia" argue that greater regional energy co-operation will not only accelerate economic and social development but also ensure peace, stability and security of the region. The paper surveys the current regional energy scenario and highlights the potential for regional energy resources to meet the growing energy demand in the region.

C. Vinodan's paper "Emerging Nuclear Order and South Asia" analyses the emerging dimensions of nuclear proliferation in South Asia. The concerns regarding nuclear security and fragile state structure in South Asia attracted global attention. The paper argues that the link between the Non-State Actors and the Pakistani nuclear establishment is posing serious challenges to the emerging regional security order in South Asia.

I place on record my deep sense of gratitude to all the contributors to this volume for their scholarly research findings which, I hope, will provide valuable insights into the contemporary scenario in South Asia vis-à-vis the emerging global order.

Kottayam **C. Vinodan**

1

Sustainable Development and Human Security: Global Context and South Asian Solutions

Sabina W. Lautensach and
Alexander K. Lautensach

The literature on sustainable development features a daunting diversity of interpretations, worldviews, conflicting moral priorities, and contrasting cultural perspectives. The differences stem mainly from diverse interpretations of development and sustainability, surrounding the two fundamental questions, what is to be sustained and what is to be developed (Khagram et al., 2003). The conflicting and much contested interpretations of sustainable development often get in the way of the arguments about its manifestations, applications and ramifications. Before we embark on any kind of argument we will clarify our particular interpretations of development and sustainability as they pertain to the geographical, political and cultural context of South Asia. Our main focus will be on India but much of what we have to say applies in principle also to Pakistan, Bangladesh, Sri Lanka, Nepal, and the smaller countries in the region.

What makes a developing country? While every country can be regarded as in a state of continuous change, in the case of those with a high percentage of citizens living in poverty that change is referred to as 'development' inasmuch as it involves policies designed mainly to counteract their poverty. Hence, for example, the United Nations' Millennium Development Goals are aimed primarily at reducing poverty [1] (UN, 2005). Beyond that vague ideal, opinions vary widely as to which deliberate changes in a country such as India in fact constitute development and which may not. It seems clear that unqualified economic growth does not fit the bill, as it neither reliably alleviates poverty (Daly and Cobb, 1994), nor do its benefits necessarily exceed its harms. [2] Hence, most definitions of development include such ideals as 'human capacity development', and social justice, and quality of life (Chen et al., 2009; Khagram et al., 2003: 116). But since abject poverty tends to function as a prime obstacle in the pursuit of those ideals, its alleviation rightfully receives top priority among the goals of development.

Even more controversial and diverse are the interpretations of sustainability. Lemons (1996: 198) defined sustainability as "the continued satisfaction of basic human physical needs, such as food, water, shelter, and of higher-level social and cultural needs, such as security, freedom, education, employment, and recreation", along with "continued productivity and functioning of ecosystems". We shall explain and expand on that definition below and merely reiterate at this point the fundamental insight that efforts to sustain ecological support structures and efforts to advance the cause of human betterment must go hand in hand; each depends on the other, so that failure on one side will prevent success on the other. As we will show below, this disqualifies economic growth, at least in its conventional conceptions, from the field of sustainable development. Most people—with the exception of neoclassical economists-understand that nothing in the physical universe, economies included, can grow indefinitely. Therefore, to pursue sustainable development other, more sustainable means of reducing poverty must be identified and promoted.

To that end, reliable methods must be available to assess and verify whether chosen policies in fact have the desired effects on sustainability and development. After two major global conferences, the 1992 United Nations Conference on Environment and Development and the 2002 World Summit on Sustainable Development (WSSD), and numerous policy and program documents arising from them (most notably Agenda 21), it seems that unanimous agreement still eludes us as to what measures might be completely sustainable. As the first objective of this essay we will present a definition of sustainable development that addresses those uncertainties, mainly through its connection with human security and with certain salient scientific principles.

Lack of definitional clarity has rarely prevented governments from using a term to promote their policies. In the wake of the WSSD, India's government formulated the following major agenda for its sustainable development (DAPL, 2002)—Combating poverty, putting people first, harnessing scientific and technological prowess, the setting of standards, institutions and legislations, conserving natural resources, and promoting sustainable development in the core sectors of the economy. Those agenda are directed towards specific targets for sustainable development (DAPL, 2002: 9). The targets are listed in Table 1.1, along with the extent to which they have been achieved to date.

We will discuss the variable extent of achievement on those targets later in this chapter. Regardless of the variable success, two important questions about the targets themselves remained unaddressed in the program documentation. Firstly, it is unclear to what extents those agenda and targets actually constitute sustainable development in terms of the definition we are about to offer. Secondly, the agenda and targets represent both ends and means to varying degrees; but even those that primarily constitute ends, such as the one about reducing poverty by 15 percent, carry an air of instrumentality —a tacit assumption that they are to lead towards more ultimate moral goals, ends of development that the vast majority of humanity subscribe to, such as an increased quality of life. Yet those ulterior goals are hardly ever spelled out as a means of justifying the targets, nor are they discussed in the various culturally contingent forms as people envision them.

Table 1.1: India's Targets for Sustainable Development and the Extent of Their Achievement

Planned Target (DAPL, 2002)	Extent of Achievement by 2011
Reduction of poverty ratio by 5 percentage points by 2007 and by 15 percentage points by 2012	In 2000, 35 percent of Indians lived on less than US$ 1 a day. In 2010, it was 21 percent. Target achieved.
All children in school by 2003; all children to complete 5 years in school by 2007	By 1999, primary enrolment was 77 percent (WRI, 2007). By 2008 it had reached 90 percent and is now achieved with delay◎approaching 100 percent in several states (World Bank, 2010).
Reduction in gender gaps in literacy and wage rates by at least 50 percent by 2007	The 2001 gender gap in literacy was 61 percent (78.3 percent vs. 47.8 percent). In 2008 it was 66 percent (73 percent vs. 48 percent) (World Bank, 2010). This indicates a gap reduction of just under 13 percent.
Increase in literacy rate to 75 percent by 2007	Achievement in 2007 was 68 percent.
Reduction in population growth between 2001 and 2011 to 16.2 percent	Between 2001 and 2011 India's population increased from 1.02 billion to 1.19 billion, equivalent to a 16.7 percent increase (Times of India, 2010). By 2009, fertility had dropped to 2.7 births almost achieved◎per woman, compared to a replacement value of 2.1.
Reduction of the infant mortality rate to 45 per 1000 live births by 2007 and to 28 by 2012	During 2000-2005, the infant mortality rate was 65 (WRI, 2007). By 2010, it had decreased to between 49 (CIA) and 53 (PRB).

Contd...

Reduction of maternal mortality rate to 2 per 1000 live births by 2007 and to 1 by 2012	From 1997 to 2005, the maternal mortality rate decreased from 6.9 to 5.5. In 2008 it was 4.5 (World Bank, 2010).
Increase in forest cover to 25 percent by 2007 and 33 percent by 2012	From 19.49 percent in 1987, forest cover increased slowly to 21.02 in 2009 (MEF, 2010).
All villages to have sustained access to potable drinking water by 2007	In 2000, 95 percent of urban and 79 percent or rural Indian households had access to potable water (WRI, 2007). By 2005 those numbers had changed to 91 percent and 94 percent, respectively (Khurana and Sen, 2008).
Cleaning of major polluted rivers by 2007 and other notified stretches by 2012	Between 1993 and 2005 the extent of freshwater pollution more than doubled; 80 percent of urban waste enters rivers unprocessed (Pepper, 2007).

Achievement data were obtained from
www.geography.about.com;www.cia.gov
www.prb.org,www.worldbank.og.in
http://www.cs.colostate.edu/~malaiya/india.html,
http://censusindia.gov.in/Census_Data_2001 unless stated otherwise.

Besides taking a critical look at the list of official agenda and targets and their achievement, addressing those deeper questions will be our other main objective in this essay. Based on the answers we will propose some conclusions regarding the potential of future sustainable development in the region.

Defining Development on the Basis of Human Security

In this section we will propose what we consider to be a more useful definition of development, one that is based on human security. In the subsequent section we will elaborate under what conditions development can also be sustainable.

Development as a term, despite its fuzziness, is used primarily in a prescriptive or evaluative sense, asserting that a particular process conforms to one's notion of progress (e.g., people moving from poverty towards affluence). In turn, the meaning of progress has been surrounded by possibly even more turbulent disputes, which on occasion gave rise to major wars. Elsewhere we suggested that one way in which those disputes can be addressed, or at least their underlying assumptions clarified, is to distinguish between the formal and the substantive concepts of progress (Lautensach, 2010). The former represents the general way in which all people formulate notions of progress, taking into account their individual ideals, values, beliefs, and aspirations.

The substantive concept then is the personalised product, what a specific individual might say when asked to explain what they mean by progress. Both concepts are prescriptive and extremely value laden; the former describes a way of thinking that seems common to human nature in general, while the latter represents the full diversity of cultural, ideological and personality-bound biases that inform people's notions of progress. The upshot is that progress cannot possibly mean the same thing for everybody, and therefore neither can development. It seems equally clear that it is those notions of progress that are shared by powerful elites that tend to inform those politically dominant ideologies that are sometimes referred to collectively as modernism (Lautensach, 2010). Modernism seeks to mask the wide diversity of culturally contingent views with a false sense of consensus, of homogeneity of purpose, under the banner of an invented single global culture (Bowers, 1993).

The tension between concerns about progress and concerns about security is as old as the history of the human species. As soon as someone expressed plans and aspirations for some form of development in concrete form, others raised questions and objections based on the possible effects (negative and positive) on security. One source of this diversity of views lies in the different ways in which people assess risks (Douglas and Wildavsky, 1983; Kasperson and Kasperson, 2001). The continuum of attributes ranges from foolhardy through daring, courageous, prudent, and cautionary, all the way to cowardly. While this diversity of subjective judgments must be regarded as a given, it does not preclude the possibility of societies to find common ground for policy consensus on acceptable risks where security challenges are concerned, nor does it diminish their obligation to try. The development of human security represents a major breakthrough as a tool for consensual risk assessment in global and regional affairs (Alkire, 2003). We shall describe briefly the ontology of human security as a concept before we apply it to define development.

Human security as a concept first surfaced in the 1994 Human Development Report of the UNDP (United Nations Development Program), in response to the growing realisation that the end of the cold war would not bring an end to armed conflict but that instead the nature of violent conflict was changing, away from traditional interstate war towards intrastate conflicts fuelled by ethnic, religious or ideological divisions. Citizens and their governments began to realise that states are not the only entities whose security ought to concern them. Regions, communities, families and individuals can only feel secure if they have reason to believe that their continued functioning is not going to be threatened at every turn.

Furthermore, the security of the state largely depends on the security of regions, communities, families and individuals. And occasionally states fail to fulfil their obligations as security guarantors, to the point of threatening the security of their own citizens. Thus, the absence of structural and personal violence (Galtung, 1969) was recognised as primary requirements for the security of the individual. These insights informed a shift in perspective from the state as the subject and object of security policy to the human individual as the centre of security considerations—from state security to human security (Griffin, 1995; UNDP, 1994). And since human beings, unlike states, are capable of sensations and emotions, human security was recognised as partly contingent on those particular states of mind that we tend to associate with human well-being. Obviously, the determinants of human security extend far beyond mere conditions for survival (Alkire, 2003).

It follows that human security depends on variables that extend beyond what has traditionally been regarded as the political arena. Descriptions of human security include a relative safety from acute infectious disease, minimum complements of safe fresh water and adequate nutrition, and a formal guarantee for basic human rights and dignity. Concern for security also became extended further into the future and applied to the future well-being of one's children, and, from middle age onward, with the well-being of their children, and so on. This long-term humanitarian concern has gradually come to inform the agenda of human security and has infused it with an attention to issues of sustainability (WCED, 1987; UN Millennium Project, 2005).

Development agencies operating under national, super-national or non-governmental umbrellas have adopted the individualist focus of human security and extended it into environmental and humanist dimensions. This re-conceptualisation is evident in several key policy documents of the United Nations. In the Secretary General's Millennium Report the UN's security agenda are defined as 'freedom from fear', and their development agenda as 'freedom from want' (United Nations, 2000). Thus, the UN's guiding principles on security are paraphrased in negative terms as freedom from a condition that is evidently undesirable. This conceptualisation of human security raised some objections (summarised in Lautensach, 2006) regarding its reliance on negative definitions and the emotive and unverifiable nature of concepts such as 'freedom', 'fear' and 'want'. Professed wants and needs can neither be readily reduced to minimum requirements for survival, nor do they seem particularly commensurable in the case of conflicting claims.

Another objection to those popular definitions states that the focus on 'freedoms' blinds the observer to the problem of limits or of scale. In any given quasi-closed system (such as an island, a desert oasis, or a planet) the extent to which the human inhabitants' needs and wants can be satisfied depends on the population size. Other variables, such as individual affluence and technological sophistication also apply, but only temporarily. For example, the same freedom from water shortage for a community in Rajasthan can be achieved without any effort for a population of a few thousand while remaining utterly unachievable if that population measures in the millions.

One way by which those objections can be addressed is to examine what sources of insecurity might threaten the global citizen. Because of the subjective nature of human security such an examination must involve consultation with the people in question. Multinational opinion surveys point towards criminal violence, armed conflicts (civil or international), terrorism, infectious disease, and 'natural disasters' as the events that people most fear. The latter include extreme weather events, climatic aberrations, pest invasions, famines, floods, land slides, earthquakes and volcanism, and meteorite impacts. Other sources of insecurity include economic collapses, personal bankruptcy, personal accidents with traumatic health effects, and chronic health problems.

By focussing directly on those sources of insecurity we eliminate some of the ambiguity and heterogeneity associated with the abovementioned 'freedoms' while gaining the advantage of focussing on more clearly defined targets (Lautensach and Lautensach, 2010). This helps facilitate proactive and preventive policy planning and enables us to enlist a host of descriptive-analytical sciences in our planning efforts. Returning to the example of water security, by focussing on possible causes of water shortage in our Rajasthan community and on the systemic requirements for local water security the observer would be forced to take into account the limits of the local system, an essential requirement for the design of long-term effective and sustainable policies.

Surveying possible sources of insecurity also allows for a comprehensive definition of human security, including four broad areas referred to as the 'four pillars' of human security (Lautensach, 2006). They include the traditional area of military/strategic security of the state; economic security, particularly as described by heterodox models of sustainable economies; health-related security, informed by epidemiology and the complex determinants of community health and health care priorities; and environmental security that models the complex interactions between human populations and their ecological support structures, i.e. the source and sink functions of their host ecosystems. The four pillars cover the same ground as the 'seven dimensions' of the 1994 Human Development Report (UNDP, 1994).

As these comprehensive models have allowed our interpretation of security to transcend the state-centred perspective, hence they can help our interpretation of development overcome purely economic or health-centred perspectives. A holistic assessment of a specific development initiative's effects on all four pillars facilitates a more realistic assessment of its benefits. This approach also calls for a re-definition of development, from the sole aim of increasing affluence or health to the holistically effective goal of enhancing human security.

Since the inception of the Four Pillar model and other similarly multidisciplinary models it has become increasingly clear that the most intriguing and challenging questions in human security deal with the interrelationships between the four pillars. Those root causes of insecurity tend to affect each other, sometimes reciprocally in a positive feedback pattern. A much discussed example is the causes of violent conflict that are often found in social injustice, economic destitution, and environmental scarcity (Homer-Dixon, 1999).

Any increase in one of those variables tends to stimulate the others, which leads to a general worsening of the situation unless drastic interventions can achieve simultaneous improvements in more than one of them. Any sustainable solution to the crisis requires improvements in all three areas of causation. The decidedly mixed success of the Millennium Development Goals program, focused primarily on economic security to the detriment of the other pillars, serves as a case in point—and so do India's mixed achievements on sustainable development (Table 1.1).

How can this multidimensional concept of human security inform a vision of more secure development? One requirement emerging from the foregoing arguments is a balanced approach across the four pillars of human security. What is to be developed, then, are the efficiency of economies and of human consumption; the capacities of societies and institutions in terms of distributive equity and social justice; the capabilities, choices and wisdom of people; and the nurturing capacity of the ecosystems we all depend on. Most significant is the inviolability of systemic limits. In other words, even the best coordinated development program will eventually encounter limit conditions which circumscribe its spatial and temporal potential. Staying within those limits without jeopardising the security of millions constitutes humanity's great challenge of sustainability.

Sustainability: Basic Requirement for Secure Development

Analyses of the diverse causes of violent conflict, such as the recent civil war calamity in Sri Lanka, underline not only the importance of interactions between the four pillars but also that their relative influence on human security may not be equally strong. They also tend to illustrate the significance of sustainability in addressing human security issues. Sustainability amounts to a balance between efforts to improve the quality of life for a human population and the continued functioning of its environmental support structures, namely ecosystems. [3] Ecosystems consist of local communities of species and their physical environment. Human populations, like all other animal populations, obtain their sustenance from ecosystems which provide food, raw materials, and energy and which recycle organic wastes back into biomass. Human populations are special in that they employ technology to maximise the benefits of those ecosystem services. But regardless of this technological windfall, the capacities of local ecosystems remain limited, which allows us to assess sustainability in rather precise terms.

Generally, the environmental impact *I* of a human population on local ecosystems is described by the *I=PAT* formula, where *P* means population size, *A* stands for the affluence or economic means per capita, and *T* represents the technological impact per capita (Ehrlich and Holdren, 1971; York et al., 2003). The maximum sustainable impact, also referred to as carrying capacity (Curry, 2006: 126) is thus described as the product of the three variables; it can be reached by small populations with a high-impact lifestyle or by larger populations where each individual demands less in terms of support services.

When a population exceeds the maximum sustainable impact it enters into overshoot, where the services of the local ecosystems are being overtaxed and, depending on their fragility, those ecosystems may undergo irreversible structural changes (Catton, 1980; McMichael, 2001; Wackernagel et al., 2002; Meadows et al., 2004). Inevitably the consequence for the population is such that various biological regulatory mechanisms lead to a decrease in population size, below the system's carrying capacity. Numerous precedents from animal populations have allowed ecologists to characterise and predict those dynamics with impressive accuracy. The major benefit of applying these basic ecological principles to human security is the clarity with which they highlight the boundary between sustainable and unsustainable collective practices. Any policy or measure that relies on unsustainable practices not only jeopardises human security, it also cannot be regarded as sustainable development.

The environmental impact can also be expressed in terms

of the area of productive land required to support a population's consumption of resources (food, energy, raw materials) and the processing of its wastes. This is referred to as that population's ecological footprint (Wackernagel and Rees, 1996). It is calculated by complex algorithms that are still being refined to take into account further pherological details (Wackernagel et al., 2002; Ewing et al., 2010). A population whose footprint exceeds the amount of accessible productive land is clearly in overshoot.

This may not always have immediate negative consequences for their security as they may obtain the shortfall from other regions that are either under populated, defenceless, or otherwise disempowered. It is, however, often unjust and supports unsustainable patterns of consumption. In order to assess whether a country is in overshoot its footprint is compared with its available bio productive land area in the fashion of an economic comparison of demand and supply (Wackernagel et al, 1997; Ronsin, Newman and Dubois, 1999; Ewing et al., 2010). The same comparison can also reveal how sustainable a purported 'sustainable development' measure actually is. We shall apply that analysis to India's targets for sustainable development below.

To summarise this sequence of causation, unsustainable practices sooner or later lead a population into overshoot, which in turn erodes environmental support structures and decreases their capacity to deliver resources and to accept wastes. This means that the environmental security of the population is increasingly threatened, which will eventually lead to shortages of food, energy, or of other commodities (conditions that contribute to 'poverty'), and in elevated levels of pollution (Homer-Dixon, 1999). Such changes invariably compromise population health and lead to economic decline, civil disorder, and vulnerability to external enemies.

Evidence is provided by the historical precedent of cultures that disappeared as a result of this sequence of effects (Diamond, 2005). The upshot is that whatever safeguards may be in place to protect the economic security of this population, its public health, its national security, and the rule of law—they seem of little help in the long term unless sustainability and environmental security are guaranteed. The same condition applies to development programs. This resonates with Barnett's (2007) finding of a mutual dependence between environmental security and peace, and it reaffirms Norman Myers' (1993) (Myers and Kent, 2004) original thesis that all security ultimately depends on environmental security.

Instead of four balanced pillars, a more appropriate metaphor would be one in which environmental security forms the basis from which the three pillars of economic, socio-political and health security support human security as a whole (Lautensach and Lautensach, 2010). For this reason we shall bypass the concepts of economic, social, and cultural sustainability. Without meaning to imply that they are not important for human security we suggest that they are either subsumed under the environmental concept (as in the case of economic sustainability) or less directly involved in the survival of individuals.

The revised model informs a new approach towards assessing the human security of countries and regions, and specifically their development efforts, at least in the long term. If human security in the long term depends first and foremost on environmental security, then the true merits of any trend or development plan can be assessed by examining the extent to which relies on, or strengthens, sustainable practices. A straightforward way to verify whether a community or country is developing sustainably is by examining the trends in that population's ecological footprint. We shall now apply this analysis to the development goals as advocated by the United Nations and by the Government of India.

Assessing the Goals of Development

The United Nations' Millennium Development Goals (MDGs) illustrate the prominent interpretation of sustainable development as advocated by governments and many NGOs at the global level. The eight MDGs are global policy targets for 2015, adopted by acclamation at the 2000 UN Millennium Summit by world leaders from 191 countries in response to urging by the US, the EU and Japan. They include ending chronic hunger, achieving universal education, gender equity, child health, maternal health, combating HIV/AIDS, environmental sustainability, and global partnership. Curiously, peace was not mentioned. The overarching goal was to "end extreme poverty worldwide" by 2015, which was defined entirely in economic terms (i.e. to cut in half the proportion of people living on less than one US dollar). Despite all good intentions and moderate progress, by the 2005 'Millennium+5' summit it became evident that the targets would not be achieved as planned. Subsequent reports (MDG Gap Task Force, 2009; UN, 2010) confirmed that assessment.

The MDG Gap Task Force (2009) proposed no conclusive explanations for the shortfall, other than vague references to the adverse influence of the global economic crisis. This interpretation reveals the program's bias towards economic growth, in terms of goal priorities, agenda, and its underlying cornucopian [4] world view. Our arguments in the preceding sections indicate that any development program is likely to fail in the long term if it weakens or fails to strengthen all four pillars of human security, for example by neglecting sustainability. The MDGs' inattention to the sustainability aspect will continue to prevent their long-term success.

The MDGs were also contingent on the assumption that population growth will halt through spontaneous demographic transition without more active pursuit of specific population targets before humanity encounters the natural mechanisms that normally regulate populations (primarily including malnutrition, aggression, and epidemics), which the projected 8-10 billion by 2050 are likely to face. Moreover, the MDGs were based on the assumption that sovereign countries can agree on how to preserve the global commons and to emancipate women, thus ignoring cultural variables and ecological exigencies. Given the evident failure of the MDG program to recognise sustainable limits, as well its other shortcomings (Amin, 2006), even its partial successes seem somewhat surprising.

The most influential reason for their failure lies in a fundamental flaw in modelling that underlies the MDGs. Hunger and poverty are generally caused by lack of food and of economic capital. Both shortages tend to occur when ecosystems lose their productivity in the course of the global environmental crisis, which manifests in ever faster resource depletion, in the effects of pollution on climate, habitat quality and public health, and in the spiralling extinction of species as ecosystems deteriorate. Five self-reinforcing causes have been identified for the crisis: economic growth, population growth, technological expansion, arms races, and growing income inequality (Coates, 1991; Daly and Cobb, 1994; Furkiss, 1974; Lautensach, 2010). It seems more than coincidental that the two main remedies prescribed by the MDG program, economic growth and technological expansion, are also two of the causes for the global environmental deterioration that in turn creates hunger and poverty. Similar concerns are raised by the partial dependence of India's sustainable development targets on economic and technological expansion.

India's development targets, listed in Table 1.1 along with their respective achievement, were drawn up in the early stages of a period of unprecedented economic growth, attributed largely to a move towards deregulation of key industries. By 2000, India's market shares in communication technology and biotechnology had skyrocketed, and the ramifications throughout its economy had allowed 130 million new consumers to leave poverty behind (Myers and Kent, 2004: 85). In the past two decades, household income has approximately doubled, and it is projected to triple in the next two (WWI, 2008). Unprecedented access for foreign investors, privatisation of industries, and deregulation of government control were cited as the main domestic factors (Chopra, 2011).

Despite the unqualified enthusiasm with which those growth figures have generally been received, they raise two fundamental concerns. Deregulation of industries and liberalisation of trade, despite their propensity to stimulate the conventional indicators of economic growth, tend to result in benefits becoming privatised while risks are nationalised; cases in point are the numerous bank bailouts by governments worldwide since the global economic downturn in 2008 and the collapse of several European governments over failing fiscal programs during the same period. In India, negative consequences are already evident in rising pollution levels and land degradation, to be followed by the inevitable effects on public health and the national economy (Singh, 2005). The second, more serious concern stems from the detrimental effect of any further economic growth on the global environmental crisis, planetary overshoot, and India's own ecological footprint. Those two concerns are not addressed in the ministerial program documents.

On the positive side, during the first decade of the 21st century, India made considerable efforts to leave the track of conventional development (which would have meant repeating the mistakes made by today's affluent countries) regarding energy, transport, building design, and food security. Renewable energy sources, such as thin-film solar photovoltaic technology, are being developed at a rate that promises to propel India into a leading position.

The growing use of feed-in tariffs as the most cost effective policy, as well as the negative publicity attracted by old-style megalomaniac power developments such as the large dams on the Narmada river (Roy, 1999), will further stimulate the growth of electricity generation from renewable sources and at smaller, community-based scales. Although India's geographic and climatic conditions render solar energy a logical top priority, the decision to pursue it must nevertheless have been met with some ideological opposition at a time when the US were officially still in denial about global climate change and opposed any concerted action towards sustainable energy solutions.

Extending what already constitutes the world's most effective railway system, the development of public transport is prioritised over individual transport, as exemplified by light rail transport projects in Hyderabad, Bengaluru, and other urban centres. Counterproductive and unsustainable developments such as the expansion of air travel and automobile market are attributable more to initiatives of private industry rather than active government policy which appear mindful of Singapore's success in de-prioritising the automobile. Sustainability is also prioritised in building design, resulting in such award-winning developments as Hyderabad's Information Technology Engineering Consultancy (HITEC) City which rated an unprecedented 56 out of 69 points on the international LEED (Leadership in Environmental Engineering and Design) scale (Singh, 2005).

The development and implementation of sustainable strategies to improve food security has become imperative for India as agricultural yields have levelled off after the boom of the 'Green Revolution' left large tracts of soil virtually devoid of micro flora and agricultural ecosystems depleted in biodiversity (WWI, 2008). Particularly promising seems the recent expansion of community supported agriculture projects, rendering more profitable organic farming practices in small hold rural enterprises, and coordinated regional regimes for water conservation (Swaminathan, 2010). Nevertheless, the prospect of continuing population growth and the uncertain impacts of climate change put in question any optimistic forecasts on food security.

These examples of significant development decisions indicate that India's development policies are guided by both a desire to learn from abroad without repeating the same mistakes, and by certain scepticism towards unfounded assumptions and harmful practices that still inform mainstream economics, such as the cornucopian delusion and the externalisation of environmental costs. This attitude is reflected in the list of official development targets, most prominently in the explicit focus on slowing population growth. India's sustainable development target was based on its 2000 National Population Policy that aimed at reducing the fertility rate to the replacement level of 2.1 by 2010.

In fourteen states this was actually achieved through improving literacy, empowerment of women, increasing the age at marriage and delaying first births (Times of India, 2010). The fact that the national average has only dropped to 2.65 (CIA, 2011) indicates that not enough action has backed up those intentions in some states. Nevertheless, while in most 'Western' countries even a public discussion of measures to slow population growth is still considered a cultural taboo (and, for politicians, political suicide), the Indian government obviously recognised that any strategy towards sustainable development would be doomed to fail without a timely reduction in population, and decided to own up to that imperative.

As for future prospects, India's development targets in the medium and long-term appear quite justified when they are individually assessed against the requirements for sustainable human security. They explicitly address the socio-political, health-related, environmental and economic pillars of human security. Certainly the educational targets of elementary schooling and closing of literacy gaps seem as realistic as they are important for political and economic security, even though achievement has been incomplete (Table 1.1). The success of the education system in Kerala, one of the poorer states with a history of Marxist governments, indicates that the achievement of educational targets is contingent neither on a rich economy nor on capitalist systems of economy and administration. Stable democratic governance distinguishes India's situation from that of most of its neighbours, greatly facilitating locally sustainable solutions.

The health-related targets of reducing the rates of infant mortality and maternal mortality seem highly commendable from a human security perspective, especially as they do not actually counteract the target of reducing population growth (Ehrlich et al., 1995). The latter is absolutely imperative in the light of increasing consumption, considering that India's 2007 ecological footprint of 0.92 gha/capita already exceeded its available bio productive capacity of 0.50 gha/capita, and the difference has increased since (GFN, 2009). In short, India is in ecological overshoot and has been for some time.

Even if through a concentrated effort on family planning and confidence in the oft-advocated demographic transition model, aided by the planned closing of gender gaps in literacy and wages, the fertility target will be achieved as seems entirely possible, this will not reverse the overshoot for decades. Lowering per capita consumption seems a doubly daunting proposition in view of the country's current economic course and its commitment to mitigate poverty, particularly the country's 170 million Dalits (WWI, 2008: 54). Therefore, its ecological overshoot is likely to worsen for some time, with inevitable consequences for the health sector and other aspects of human security. Not surprisingly, the health targets have not been achieved (Table 1.1).

In the light of the foregoing the prospects for the environmental targets appear somewhat mixed as well. Even assuming that the increase in forest cover means an actual increase in forest ecosystems, as opposed to tree monocultures such as the ones widely developed, for example, in Malaysia, the increased demand in ecosystem services is unlikely to be met. Especially the increased demand for fresh water will not be sustained from existing aquifers, on which 85 percent of the population depend (Khurana and Sen, 2008), particularly in view of possible climate change. Some additional latitude in that respect could be provided by the targeted cleanup of major rivers, which, however has not been achieved (Pepper, 2007).

Current data on river pollution are only one indicator among many that give rise to concern over a lack of awareness for sustainability and the influence of the modern growth ideology in the growing middle class. This is evident in the unqualified praise frequently expressed in the media for the removal of "archaic" government regulation of economic activities (Chopra, 2011), while the increasing water pollution is blamed by others on "poor government oversight" (Pepper, 2007). Clearly the essential role of ecosystem services in making sustainable development possible is being neglected and underemphasised amidst the ubiquitous exuberance about current growth statistics.

It is those counterproductive ideological currents and imported short-sighted modern values that lie at the heart of the problem. Because of India's overshoot, further increases in consumption would have to be met primarily by increased imports from other countries which are themselves likely to be in increased danger of overshoot. On a planet which collectively consumes more than 140 percent of what its support structures can sustain (WWF, 2010), no country can expect to further increase its consumption without incurring the risk of recrimination from other countries whose shares might be correspondingly diminished.

Moreover, the collective planetary overshoot worsens, decreasing the resources available to future generations, which also will be more populous. This central problem was ignored by the authors of the MDGs, and it continues to prevent the global success they envisioned. Its negative effect on India's aspirations for sustainable development seems similarly unavoidable. Certainly once global trade begins to suffer from the inevitable scarcity of cheap fossil fuels, South Asians will be hard pressed to return to a more self sufficient existence. Even India's democratic structure of governance will not help unless people's values change (Homer-Dixon, 2006: 306). To return to Khagram's (2003) question as to what is to be developed: the answer is, primarily, wisdom.

India is not alone in its unfortunate state of overshoot. Its neighbours in the region fare hardly better as illustrated by the comparisons of their footprints and bio-productive capacities in Table 1.2. India, Bangladesh, and Pakistan show similar extents of overshoot, indicating that they import products of bio-capacity from other countries outside the region. Nepal and Sri Lanka show more severe overshoot, indicating an even greater dependence on outside help. The trends associated with those numbers, derived from data on their total bio-capacities over the past decades, suggest that overshoot will worsen for all countries except for Pakistan. In Pakistan, however, the projected population increase is higher than for all other countries (probably linked to a deficit in the empowerment of women) which will cause additional hardship there.

Table 1.2: Comparison of the Available Bio-productive Capacities and Claimed Ecological Footprints of the Major South Asian Countries*

Country	Bio-productive Capacity per Person (=available footprint) [gha] and trend (GFN, 2009)	Claimed Footprint per Person [gha] and trend (GFN, 2009)	Projected Population Increase 2009-2050 (PRB, 2010)
Bangladesh	1.37 steady	1.62 rising	37 percent
India	1. 50 decreasing	1.92 steady	49 percent
Nepal	1.6 decreasing	1.6 decreasing	67 percent
Pakistan	1.42 decreasing	1.76 decreasing	85 percent
Sri Lanka	0.43 decreasing	1.20 steady	24 percent

*both are given in global hectares (gha) per person, a normalised unit of land area. The projected population increase for each country allows an estimation how rapidly a country's capacity is likely to drop.

India's economic target of reducing poverty parallels the first of the MDGs, and it has shown impressive achievement in that regard (Table 1.1). However, by itself the target appears questionable unless it is backed up by explicit plans how the correspondingly increased consumption is to be sustained. Optimistic projections of a hunger-free population of 1.4

billion in 2020 are supported only by vague references to hypothetical increases in agricultural production (Swaminathan, 2010). Its population will continue to grow for several decades, even if India attains the replacement fertility of 2.1 births/woman, because present population growth carries considerable momentum (Ehrlich and Ehrlich, 1995). Those further population increases and the demands they will incur are rarely discussed by conventional development advocates. The kind of drastic redistribution of assets that might sustain those demands in food alone seems highly unlikely in view of the current ideological climate.

To summarise, several observations contribute to this sceptical view about South Asia developmental aspirations, and India's in particular, and about the optimism that the current economic boom has so widely inspired. First, there is the theoretical uncertainty of any development goal based on the assumption of continuing economic growth (Daly and Cobb, 1994). Secondly, India's overshoot and other human security indicators, and those of its neighbours, suggest a deepening dependency on bio-productive support from outside the region. Thirdly, many existing major development policies (for example, those pursued by the World Bank) rely on measures, such as stimulating economic growth, that are likely to further contribute to the very problems they seek to address. In contrast, examples of smart development receive too little support.

Even so, the prospects for the region are probably not as bleak as this summary makes them out. We take particular encouragement from the prominent role played by values, ideals, cultural beliefs and spiritual resources in connection with those problems (Lautensach, 2010). Just as many of the causes lie in the shallow values and ideals of the neo-liberal ideology of growth that recently arrived on the subcontinent, it is equally true that in South Asia they form a very thin veneer on top of a rich trove of cultural traditions much older than anything the West can offer.

Surely, four millennia of civilised history amount to sufficient cultural resources and spiritual resilience to provide an ample basis for catagenesis (Homer-Dixon, 2006: 300), a renewal after a breakdown. That philosophical and cultural heritage has on occasion been synthesised with Western ideas about sustainable development into some powerful symbols of hope. They include urban solutions such as the experimental city of Auroville, and rural schemes such as the bio-villages established by the M.S. Swaminathan Research Foundation (MSSRF), and many other initiatives at small and large scales. The small Himalayan nation of Bhutan has single-handedly set a new standard by reconceptualising its sustainable development under the new paradigm of Gross National Happiness (Wells-Bruges and Wells-Bruges, 2001). Even if the current situation makes sustainability appear out of reach, the cultural capital and human potential to achieve it have always been close to the surface in South Asian peoples.

End Notes

1. The UN defined individual poverty as earning less than the equivalent of one US dollar per day (United Nations, 2000). Despite its widespread acceptance, this definition seems unsatisfactory for at least two reasons. First, the extent to which the currency equivalent of one dollar can buy essential goods and services varies considerably around the world. A fixed amount of money therefore cannot define the same condition of living worldwide. Secondly, it says little about the actual quality of life as the individual perceives it in his/her cultural context. The gross national happiness policies pursued by the government of Bhutan represent a reasonable alternative approach, and others have been proposed.
2. As a case in point might serve the European Industrial Revolution, which gave rise to vast impoverished urban proletariats and, by the time standards of living were finally raised, contributed to a global environmental crisis of unprecedented magnitude that may well spell the end of all 'development' (Odum and Odum, 2001). More specifically, industrial/environmental disasters such as the 1984 Bhopal

incident show that some manifestation of economic growth by no means qualify as development, and many others seem ineffective (Bahuguna, 2002).

3. Because of its essential role for human survival, we refer to sustainability mainly in its environmental meaning, according to Lemons' (1996: 198) definition quoted at the onset of this chapter. We shall briefly discuss economic, social, and cultural sustainability later in this section. We regard the popular 'Brundtland' definition of sustainability (WCED, 1987) to be of little use because of its lack of conciseness, inattention to meta-ethical considerations and its neglect of fundamental ecological limitations. A more useful definition, attributed to Steve Goldfinger (Chambers et al., 2000: 2), states that a sustainable community is one that converts resources into waste no faster than ecological support structures can convert the waste back into resources.
4. Cornucopianism refers to the belief that the growth of economies, and of human populations, is not subject to any physical limits (Ehrlich and Holdren, 1971). Despite its lack of scientific basis it has long served as an ideological mainstay of neoclassical economics.

References

Alkire, Sabina (2003): "Concepts of Human Security", In *Human Insecurity in a Global World,* Chen, L., Fukuda-Parr, S. and E. Seidensticker Eds., Cambridge, Mass: Asia Centre, Harvard University, pp.15-39.

Amin, Samir (2006): "The Millennium Development Goals: A Critique from the South". *Monthly Review,* March http://www.monthlyreview.org/0306amin.php.

Bahuguna, Sunderlal (2002): "Sustainable Development in India: Perspectives", In *Sustainable Development: Learnings and Perspectives from India,* Briefing Document on the World Summit on Sustainable Development, Johannesburg. Government of India: Ministry of Environment and Forests. http://envfor.nic.in/divisions/ic/wssd/doc4/consul_book_persp.pdf.

Barnett, J. (2007): "Environmental Security and Peace", *Journal of Human Security*, 3(1): 4-16.

Bowers, Chet A. (1993): *Education, Cultural Myths and the*

Ecological Crisis: Towards Deep Changes, Albany, NY: State University of New York Press.

Catton, W.R. Jr. (1980): *Overshoot: The Ecological Basis of Revolutionary Change,* Urbana: University of Illinois Press.

Chambers, N., Simmons C. and M. Wackernagel (2000): *Sharing Nature's Interest: Ecological Footprints as an Indicator of Sustainability*, London: Earthscan Publishing Ltd.

CIA (Central Intelligence Agency) (2011): *The World Factbook* https://www.cia.gov/library/publications/the-world-factbook/geos/in.html.

Chen, L., Fukuda-Parr, S. and E. Seidensticker Eds., (2003): *Human Insecurity in a Global World,* Cambridge, Mass: Asia Centre, Harvard University.

Chopra, Anuj (2011): "Liberalisation brings India billionaires and Maoists", *The National,* http://www.thenational.ae/thenationalconversation/industry-insights/economics/

Coates, Joseph F. (1991): "The sixteen sources of environmental problems in the 21st century", *Technological Forecasting and Social Change* 40: 87-91.

Curry, P. (2006): *Ecological Ethics: An Introduction*, Cambridge, UK: Polity Press.

Daly, H.E., and J.B. Cobb (1994): *For the Common Good: Redirecting the Economy towards Community, the Environment, and a Sustainable Future,* Boston: Beacon Press, 2nd ed.

DAPL (Development Alliance Pvt. Ltd.) (2002): "Empowering People for Sustainable Development", *Government of India, Ministry of Environment and Forests,* http://moef.nic.in/divisions/ic/wssd/doc1/main.htm.

Diamond, Jared (2005): *Collapse: How Societies Choose to Fail or Succeed,* London: Viking Penguin.

Douglas, Mary and Aaron Wildavsky (1983): *Risk and Culture,* Berkeley, CA: University of California Press.

Ehrlich, Paul R. and J. Holdren (1971): "The Impact of Population Growth", *Science* 171: 1212-1217.

Ehrlich, Paul R., Ehrlich, A.H. and G.C. Daily (1995): *The Stork and the Plow,* New York: G.P. Putnam's Sons.

Ewing, B., Moore, D., Goldfinger, S., Oursler, A., Reed, A. and M. Wackernagel (2010): *The Ecological Footprint Atlas 2010.* Oakland, CA: Global Footprint Network.

http://www.footprintnetwork.org

Furkiss, Victor (1974): *The Future of Technological Civilisation,* New York: Braziller.

Galtung, Johan (1969): "Violence, Peace, and Peace Research", *Journal of Peace Research,* 6 (3): 167-191.

GFN (Global Foorprint Network) (2009): "Footprints for Nations", http://www.footprintnetwork.org/en/index.php/GFN/

Griffin, Keith (1995): "Global Prospects for Development and Human Security", *Canadian Journal of Development Studies,* XVI (3): 359-370.

Homer-Dixon, Thomas (1999): *Environment, Scarcity, and Violence,* Princeton: Princeton University Press.

Homer-Dixon, Thomas (2006): *The Upside of Down: Catastrophe, Creativity, and the Renewal of Civilisation, Washington,* DC: Island Press.

Hughes, B.B., Irfan, M.T., Khan, H., Kumar, K.B., Rothman, D.S. and J.R. Solorzano (eds.), (2009): *Patterns of Potential Human Progress. Vol.1: Reducing Global Poverty,* New Delhi: Oxford University Press India.

Kasperson, Jeanne X. and Roger Kasperson (eds.), (2001): *Global Environmental Risk,* New York: United Nations University Press.

Khagram, S., Clark, W.C. and D.F. Raad (2003): "From the Environment and Human Security to Sustainable Security and Development", In *Human Insecurity in a Global World,* Chen, L., Fukuda-Parr, S. and E. Seidensticker (eds.), Cambridge, Mass: Asia Centre, Harvard University. 107-135.

Khurana, Indira and Romit Sen (2008): "Drinking Water Quality in Rural India: Issues and Approaches", http://www.corecentre.co.in/Database/Docs/DocFiles/drinking_water.pdf.

Lautensach, A.K. (2006): "Expanding Human Security", *Journal of Human Security,* 2(3): 5-14.

Lautensach, A.K. (2010): *Environmental Ethics for the Future: Rethinking Education to Achieve Sustainability,* Germany: Lambert Academic Publishing.

Lautensach, A.K. and S.W. Lautensach (2010): "Prioritising the Variables Affecting Human Security in South East Asia", *Austrian Society for South-East Asia Journal,* 3(2).

Lemons, John (1996): "Afterword: University education in

sustainable development and environmental protection", In J.B. Callicott F.J.R. da Rocha (eds.), *Earth Summit Ethics: Towards a Reconstructive Post-modern Philosophy of Environmental Education* (pp. 193-217), Albany, New York: State University of New York Press.

MAB (Millennium Assessment Board) (2005): *Millennium Ecosystems and Human Wellbeing,* Washington DC: Island Press.
http://www.millenniumassessment.org//en/Products.Synthesis.aspx.

MDG (Millennium Development Goals) Gap Task Force (2009): *Strengthening the Partnership for Global Development in a Time of Crisis,* New York: United Nations.

McMichael, A.J. (2001): *Human Frontiers, Environments and Disease: Past Patterns, Uncertain Futures,* Cambridge: Cambridge University Press.

Meadows, D., Randers, J. and D. Meadows (2004): *Limits to Growth: The 30-Year Update,* White River Junction, VT: Chelsea Green Publishing Co.

MoEF (Ministry of Environment and Forests) (2010): *Role and Mandate of the Ministry: Annual Report 2009-10,* New Delhi: Government of India, Ministry of Environment and Forests.

Myers, Norman (1993): *Ultimate Security: The Environmental Basis of Political Stability,* New York: W.W. Norton and Co.

Myers, Norman and Jennifer Kent (2004): *The New Consumers: The Influence of Affluence on the Environment,* London: Island Press.

Odum, Howard T. and Elisabeth C. Odum (2001): *A Prosperous Way Down.* Boulder, CO: University Press of Colorado, cited by Orr 2005 about his challenges of Sustainability.

Pepper, Daniel (2007): "Choking on Pollution in India: Dead Rivers and Raw Sewage", *Der Spiegel* (July 6).
http://www.spiegel.de/international/world/0,1518,493033,00.html.

PRB (Population Resource Bureau) (2010): *World Population Datasheet 2010.* http://www.prb.org/pdf10/10wpds_eng.pdf.

Rees, William (2004): "Waking the Sleepwalkers–Globalisation and Sustainability: Conflict or Convergence", In *the Human Ecological Footprint,* Guelph, Canada: University of Guelph. (pp. 1-34).

Roy, Arundhati (1999): *The Greater Common Good,* Mumbai: India Book Distributor.

Sachs, Jeffrey D. and J. W. McArthur (2005): "The Millennium Project: A plan for meeting the Millennium Development Goals". *The Lancet 365*: 347-353.

Singh, Amita (2005): *Administrative Reforms: Towards Sustainable Practices,* New Delhi: Sage Publications.

Swaminathan, M.S. (2010): "We need profits, passion in farming", *Hindustan Times* (30 December), http://www.hindustantimes.com/StoryPage/Print/644727.aspx

Times of India (2010): "Azad Rules out Law to control Population Growth", (6 May) http://timesofindia.indiatimes.com/india/Azad-rules-out-law-to-control-population-growth/articleshow/5895087.cms.

UNDP (United Nations Development Programme) (1994): *Human Development Report: New Dimensions of Human Security,* New York: UNDP.

UN Millennium Project (2005): *Investing in Development: A Practical Plan to Achieve the Millennium Development Goals,* New York: United Nations.

United Nations (2000): *We the Peoples: The Role of the United Nations in the 21st Century,* New York: United Nations, www.un.org/millennium/sg/report/key.htm.

United Nations (2009): *End Poverty 2015: Millennium Development Goals,* New York: United Nations, http://www.un.org/millenniumgoals/

United Nations (2010): *The Millennium Development Goals Report 2010,* New York: United Nations.

Vitousek, P.M. Mooney, H.A. Lubchenko, J. and J.M. Melillo (1997): "Human Domination of Earth's Ecosystems", *Science* 277: 494-499.

WCED (World Commission on Environment and Development) (1987): *Our Common Future: The Brundtland Report*, Oxford: Oxford University Press.

Wackernagel, Mathis and William Rees (1996): *Our Ecological Footprint: Reducing human impact on the earth,* Oxford: John Carpenter.

Wackernagel, M., Schulz, N.B., Deumling, D., Linares, A.C., Jenkins, M., Kapos, V., Monfreda, C., Loh, J., Myers, N., Norgaards, R., and J. Randers (2002): "Tracking the ecological overshoot of the human economy", *Proceedings of the National Academy of Sciences (USA),* Vol. 99, pp. 9266-9271, http://www.pnas.org/cgi/reprint/99/14/9266.pdf.

Wells-Bruges, M. and J. Wells-Bruges (2001): "Land of the Thunder Dragon", *Resurgence 206* (May/June): 22-23.

World Bank (2010): *India Country Overview 2010,* http://www.worldbank.og.in.

WWI (Worldwatch Institute) (2008): *State of the World: Innovations for a Sustainable Economy, New* York: W.W. Norton.

WRI (World Resources Institute) (2007): *Earth Trends: The Environmental Information Portal.* http://earthtrends.wri.org/pdf_library/country_profiles/pop_cou_356.pdf.

WWF (World Wildlife Fund) (2010): *Living Planet Report 2010.* http://wwf.panda.org/about_our_earth/all_publications/living_planet_report/

York, R., Rosa, E.A., and T. Dietz (2003): STIRPAT, IPAT and Impact: Analytic Tools for Unpacking the Driving Forces of Environmental Impacts", *Ecological Economics* 46(3): 351-365.

2

Social Capital and Region Building in South Asia

Mohanan B. Pillai and P.M. Joshy

This paper is an attempt to critically examine the notion of social capital in the contexts of the ascendant neo-liberal global regime and South Asian regional integration. The neo-liberal agencies have been trying to transpose the Third World State in the developmental thinking and the remaining space has been occupied by the Western aided NGOs. Through this new version of civil society, there is a concerted effort to transcend the rigid cultural compartments in order to integrate every nook and corner of the world into the market logic of capitalism.

According to the neo-liberal line of thinking the engagements in the civil society produce social capital. However, in this paper we argue that the same process also generate many anti-democratic developments in the society. A discussion about South Asian integration is incomplete without an understanding of the politics of identities. In the first part of the study we critically analyse the notion of social capital propounded by the neo-liberal agencies and the second part deals with identities in the context of nation-building in South Asia and consequently examine some critical aspects of the notion of social capital in the region.

Social Capital: A Conceptual Note

The notion of social capital is one of the most debated concepts in the context of democracy and development. The dissemination of Robert D. Putnam's version of 'social capital' through the global capitalist institutions like the World

Bank has opened up a series of debates about the concept in the context of the new developmental strategies under neo-liberalism. The World Bank viewed social capital as the "missing-link" in development (see World Bank on social capital). The critical studies by John Harriss in his book entitled, "Depoliticizing Development: The World Bank and Social Capital" came to the conclusion that the very idea of social capital forwarded by the World Bank is an "anti-politics machine." The Bank has been trying to "depoliticise" the development by transcending the class divisions.

However, the concept was discussed much before in critical context by the thinkers of different theoretical persuasions like Pierre Bourdieu. Bourdieu, a French cultural theorist, in his critical investigation on the question of how the social classes are reproduced through the construction of meanings has come with the idea of 'connections' in the society which yielded social capital and the dominant classes convert it into economic capital, consequently, the sustained domination of the upper class. In this point of view the social capital is not the characterised feature of the society as a whole, but it is an aspect of 'differentiation of classes' ultimately 'an instrument of power' (Harriss, 2001: 5).

However, Bourdieu's notion of social capital was sidelined in the academic and policy discussions and an American sociologist James S. Coleman, a 'rational choice' theorist, became so acceptable to the neo-classical economists who have had profound influence on the World Bank (Harriss, 2001: 6). In his celebrated work "Social Capital in the Creation of Human Capital," Coleman viewed that the social capital exists in the structure of relations between the individuals. It does not reside in the individuals (Coleman, 1998: 95-120). Interestingly, on the contrary to the understanding of these two thinkers on Social Capital, Robert D. Putnam revived the concept by creating new meanings and it emerged as a fashionable concept among the social scientists and policy makers.

Social capital addresses issues like the 'variations in government performance' and the 'levels of socio-economic development' in various societies (Harriss, 2001: 6-7). Putnam described social capital in terms of 'civic engagement' and the meaning of the concept social capital has changed from the understandings of Bourdieu and Coleman, and it became a testimony for the 'property of a whole society' (Harriss, 2001: 7). The World Bank has been promoting Putnamian notion of social capital through its 'new developmental strategies.' The conventional meaning of the concept has changed and now it is imaged in a new vast canvas which cut across the conflicting classes and interests. Under neo-liberal global regime, by promulgating a new version of Social Capital, the Bank has been weaving an 'apolitical space,' for easy integration and extraction of the Third World economy.

According to Robert D. Putnam: "Social capital…refers to features of social organisation, such as trust, norms and networks that can improve the efficiency of society by facilitating coordinated actions" (Putnam, 1993: 35-42). The interactions through the networks shifted the participants' sense of self from 'I' to 'We' and that resulted in search for a more co-operational line of thinking for achieving mutual benefit (Putnam, 1993). Social capital is also refers to 'civic engagement,' and it is measured through analysing the political participation within a community, newspaper readership, the density of the voluntary associations. This 'civic engagement' creates social capital. The dense network in a community brings forth more cooperation among the citizens for mutual benefit (Putnam, 1993: 173).

In his analysis on America Putnam argued that the participations in political activities declined considerably in US and the people are showing greater importance to participate in a new generation of voluntary organisations as compared to the traditional form of civic organisations (Putnam, 1993: 4-6). In this background it is very clear that under neo-liberal global regime there are attempts to globalise the American version of socio-political models across the world. When we look through the traditional liberal discourses, it is evident that the State and civil society are playing mutually constitutive role to protect the rights of the citizen. However, under neo-liberalism the positive role of the State has been undermined.

Under neo-liberalism, there is a concerted effort to transpose the Third World State from the developmental thinking. In the name of good governance, radical democracy, grassroots development and people's participation, a large array of western aided NGO network is being deployed as the "civil society." Through this new version of "civil society" that the neo-liberal policies have been implemented across the World. Paradoxically, the shrinking state has also been too much militaristic in implementing laws and treaties of the market.

Under neo-liberal theory, 'the states are proposed to use its monopoly of the means of violence to preserve the freedom of the individuals, businesses and corporations to operate within this institutional framework of free markets and free trade, at all costs'(Harvey, 2005: 64). The neo-liberal State has reduced to mere facilitator for the global finance capital and the civil society has been "hijacked" by the neo-liberal agencies. The active participation in the civil society is viewed as the only way to develop social capital. This is an attempt to surpass the antagonistic cultural compartments in order to integrate and extract the Third World economy. The capitalist knowledge industry has been entertaining stylish and apolitical developmental patterns, but it is not exempted from criticisms.

The Putnamian notion of social capital was criticised at different angles. Social capital also may produce ethnic strives. This stream of thought emphasised that a particular kind of civic engagement generated ethnic conflicts. This is so evident in the multi-ethnic societies like South Asia, because the very logic of societal composition is very much suitable for disintegration rather than integration. The networks across, means the "bridging" of different cultural communities, may brings forth cooperation and development, but too much "bonding" within the cultural community may produce anti-social and anti-developmental results.

The ascendancy of the Hindu Right in India is an example (Varshney, 2001; Varshney, 2002; Chandhoke, 2003). The Hindu Right forces, generally the Sangh Parivar has been penetrating to the society through a series of organisations working in the civil society. These are the organisations engaging in the spheres of education and other "social empowerment" programmes. These institutional structures have been generating social capital, but not entertaining a democratic dialogue between the 'Self' and the 'non-Self.' Another critical approach by John Harriss viewed that the very notion of social capital inherited social exclusion. In social engagement particularistic interests may be get prominence and the common social good may be side lined. In contrast to the social mobilisation and political organisation the civic engagement will not produce public good.

Here, he stressed the importance of political struggle for attaining a more equitable social order. So in this background it can be argued that the depoliticized developmental pattern propagated by the World Bank and other global capitalist agencies have clear-cut agenda, transcending social antagonism and the better execution of the neo-liberal policies across the world. Coincidently, the new developmental strategies also strengthen anti-democratic forces in the society.

Anatomy of South Asia

The South Asia is a region of rich diversities and complexities in which ethnicity, religion, caste, language are so keen in determining the nature of politics; and the configuration of these multiplicities most often precipitates political polarizations and mobilizations which quite often lead to violent conflicts among different communities of the region that cut across state borders. Generally speaking, these conflicts are a modern phenomenon and are intertwined with the logic of nation-building and interest articulation within a competitive capitalist system (Joshy and Mohanan, 2010).

In South Asia nation-building is progressing on a majoritarian line. The basic thread is the protection of the 'Self' by distancing 'Other.' In this process the minorities are excluded and most probably they are portrayed as a security threat to the Nation. In order to understand these complex relations, the traditional Realist paradigm in IR is not sufficient. In this section of the article we discuss the Constructivist line of thinking on identity and security in the context of nation-building in South Asia.

The question of 'representation' is one aspect that was totally discarded by the traditional IR theorists. In their critical endeavour, the politics of identity is an important contribution by the critical constructivists (Katzenstein, 1996). For the critical constructivists, objects of security and insecurities are mutually constituted in a variety of ways that may privilege a certain conception of identity over others. Operating within a framework of meanings, assumptions, and distinctive social identities, the representation of the 'Other,' their identities and what constitutes insecurity 'imaginaries' are left open to the dynamics of interpretation, whereby relations of identity may also be produced, enforced, and reified in a conflictual manner (Muppidi, 1999).

Further, construction of identities influencing security dynamics may not simply be confined to rigid interstate dynamics, but may also be mediated by "complex network of social relations, cultural traditions, and political structures..." involving state security elites themselves (Niva, 1999). Thus, critical constructivists assume that all social (in) securities are culturally produced (Weldes, 1999: 1). It is the "cultural process through which insecurities of states and communities...are produced, reproduced, and transformed" (Weldes, 1999: 2).

Security is inextricably interlinked with identity politics. How we define ourselves depends on how we represent others. Representations of the 'Other' as a source of danger to the security of the 'Self' in conventional understandings of security are accompanied by the dehumanisation and stereotyping of the 'Other.' The 'Other' gets reduced to being a danger and hence an object that is fit for surveillance, control, policing and possibly extermination. This logic of the discourse of security dictates that the security of the 'Self' facilitates and even demands the use of policing and violence against the 'Other' (Anand, 2005: 203-215). Central to the concept of postcolonial insecurity is what Sankaran Krishna calls postcolonial "anxiety", defined as an ideological drive of postcolonial state leaders to achieve successfully the "modern enterprise of nation-building" (Krishna, 1999: xvii-xix).

According to him the process of nation-building makes its 'Others' not only in its 'cartographical limits' but also beyond it (Krishna, 1994: 507-521). Interestingly, the central to the metaphor of creating a nation as something "ever in the making but never quite reached," is the idea of nationalism (Das, 2002). The unfinished project of nation-building in South Asia generates so many 'Others' both within and out side the Statist territorial boundaries.

In the preface of his book Adeel Khan said: "Nationalism is a way of dealing with a world where 'everything melts in the thin air': It is a form of self-love in which individuals celebrate their group identity, and like all forms of self-love gets its strength more from its hatred of others than its love for its self." He continues, "Nationalism is probably the only form of self-love that gets its life-blood from an institution, the modern state" (Khan, 2005). The modern nation-State is the embodied form of nationalism which is characterised by its homogeneous existence. As far as South Asia is concerned the nationalism is border-crossings, reciprocity of opposites.

The policies of the colonial masters integrated the nations economically and also the nature of mobilisations against colonialism by the native political elites determined the cartographical setting of South Asia. The 'cutting and shaping' has progressed on a majoritarian line and in the postcolonial period to a greater extend the painful history of the past has been manipulating for distancing the 'Other' from the 'Self.' In the postcolonial period the old political practices got transformed into more sophisticated mobilisations and manoeuvrings (Katzenstein, 1996).

However, the nation-building process in a multicultural society like South Asia has been progressing on a majoritarian line. The notion of security is the running force and by creating 'Others' both within and outside the territorial parameters of the nation, the postcolonial state has been perpetuating an 'insecurity syndrome,' which dehumanise the Other by generating a chain of images through various mediums which include the national rituals commemorating various epochs in the history of the 'nation,' Cricket matches (Krishna, 1993) and other regional engagements. War is also a tool which provides vast canvas for caricaturing the 'Other' and strengthening the 'Self.' In a logical progression of insiders/outsiders, purity/pollution, civilised/uncivilised the ruling class has been sustaining an enemy image about the 'Other.' This seems to be a major hurdle for South Asian integration.

Social Capital: South Asian Experience

The liberal discourses generally viewed the civil society an essential prerequisite for democracy and development. However, under neo-liberal globalism the conventional meaning of civil society has changed and replaced the State in the name of good governance, participatory democracy etc. During the 1990s, the Third World countries were forced to integrate their economy and polity in accordance with the interests of global finance capital. Then onwards, South Asia is also experiencing a new version of civil society. This is an attempt to transcend the conflicting classes.

As far as South Asia is concerned the attempts for regional

integration is not a new one. Historically, there are so many problems withstanding against such endeavours. The logical existence of the South Asian states as independent nation-states (we have already discussed), the hegemony of India in the region, bilateral problems etc are some of the issues which resist integration.

In South Asia, civil society is a complex terrain of activity in which ethnic, religious and linguistic groups etc are playing vital roles. The post colonial elites of South Asia attempted to transcend the rigid cultural divisions with the Western rationalist discourses. The notions like the 'national security,' 'national interest,' 'national development' were become the techniques to 'silence' the dissent voices. However, under neo-liberal global regime, the changed State-society relations spawned many resistance movements and identity assertions across South Asia.

The "failed" proposals of modernity prompted the marginalised sections to struggle against the State as well as the unbridled pursuit of the global finance capital. The critical activities of those excluded sections challenged the conventional understanding of society and polity. In this background there are majority identity assertions like the Hindu Right in India. The Hindutva, the ideology of the Hindu Right has a totalising imperative because it owes its energy from the Western rationalist discourses. So there are attempts to 'silence' the critical social engagements in various ways and which are aimed to smoothen the path of the financial globalisation.

The new developmental proposals by the global financial agencies strengthen the argument that the active participation in the "civil society" generates social capital. The secular and all inclusive institutions in the civil society are a boon for democracy, but the communal organisations are also working through the civil society. They also generate social capital through informal networks. For example, the Hindu Right organisations are so influential in the Indian civil society. They are generating an insecurity complex through various mobilisations in the civil society and the notion of 'security' is a critical factor as concern with the Hindu Right forces (Joshy and Seethi).

Their organizational net-works include a large array of institutions like Vanavasi Kalyan Ashram (founded in 1952) working among the Adivasis for re/conversion, Vidya Bharati (1977) mainly in the education sector, which is having 20933 educational institutions, Vivekananda Kendra (1963) spreading Vivekananda's teachings, Vishwas Hindu Parishad (1963) to counter the secular values and unite the different Hindu sects , Sanskar Bharati, to promote Hindu awareness, Seva Bharati (1990) for those Hindus who are not formally a part of the Sang family, Hindu Jagaran Manch (1990) etc. The most important fact is that the communal organisations are giving no room for a friendly dialogue between the 'Self' and the 'non-Self.'

The majority ethnic group has profound influence in the State apparatus. The case of Punjabi ethnic community in Pakistan and Hindu Right forces during the NDA regime in India are examples. Their ideology has been produced and reproduced through the Statist machineries as well as the civil society. In this spectrum, India is regarded as the 'Other' of the Pakistani 'Self' and vice versa. In this scenario, regional integration may bring forth legitimacy crisis to the ruling class.

Conclusion

On the one hand, through the new development strategies the neo-liberal agencies have been trying to transcend the class antagonism in order to facilitate the process of financial globalization, on the other many anti-democratic forces have also been channelized through the same track. In this scenario, the civil society and social capital are so critical, because the very process enabled the neo-liberal agencies to appropriate the Third World space for capitalist expansion and at the same time it generates social conflicts and ultimately produces an insecurity complex in the region. The fundamentalist religious forces have been penetrating into the society through the informal networks and generating social capital. In other words those forces are not allowing space for democratic expansion. In short, the strengthening of the social forces thwart the attempts of integration and the global capitalist agencies have been trying to bypass such strict societal divisions through new strategies. The net outcome is widening social divisions and social conflicts across the region.

References

Anand, Dibeysh (2005): "The Violence of Security: Hindu Nationalism and the Politics of Representing 'the Muslim' as a Danger," *The Round Table,* 94 (379): 215.

Chandhoke, Neera (2003): *The Conceits of Civil Society*, New Delhi: Oxford University Press.

Coleman, James S. (1998): "Social Capital in the Creation of Human Capital", *The American Journal of Sociology,* 94: 95-120.

Das, Runa (2002): "Engendering Post-Colonial Nuclear Policies through the Lens of Hindutva: Rethinking the Security Paradigm of India," *Comparative Studies of South Asia, Africa and the Middle East*, XXII (1&2).

Harriss, John (2001): *Depoliticizing Development: The World Bank and Social Capital*, New Delhi: Leftword Books, p. 5.

Harvey, David (2005): *A Brief History of Neo-liberalism*, New York: Oxford University Press, p. 64.

Joshy, P.M. and Pillai, Mohanan B (2010): "Nation-Building and Foreign Policy Behaviour of India in the Regional setting of South Asia," in Mohanan B. Pillai and L. Premashekhara, (ed.), *India's Foreign Policy: Continuity and Change*, New Delhi: New Century Publications.

Joshy, P.M. (2010): "Perils of 'Security': Identity and Nation-Building in South Asia with a Special Reference to India and Pakistan", *International Journal of South Asian Studies*, 4 (2).

Joshy, P.M. and K.M Seethi (2011): "Interrogating Security: The Hindu Right and the Nuclear Question", *South Asian Journal of Diplomacy,* 1(1): 163-75.

Katzenstein, Peter (ed.), (1996): *The Culture of National Security: Norms and identity in World Politics,* New York: Columbia

University Press.

Khan, Adeel (2005): *Politics of identity: Ethnic nationalism and the state in Pakistan,* Sage Publications, New Delhi.

Krishna, Sankaran (1993): "Cricket, Nationalism and the Sub-continental Fan," *Pravada,* No. 2.

Krishna, Sankaran (1994): "Cartographical Anxiety: Mapping the body politic in India," *Alternatives,* 19: 507-521.

Krishna, Sankaran (1999): *Post-Colonial Insecurities: India, Sri Lanka and the Question of Nationhood,* University of Minnesota Press, Minneapolis, pp. xvii-xix.

Muppidi, Himadeep (1999): "Post-Coloniality and the Production of International Insecurity," in Jutta Weldes (ed.), *Cultures of Insecurity: States, Communities, and The Production of Danger*, Minneapolis: University of Minneapolis, p.124.

Niva, Steve (1999): "Contested Sovereignties and Post-Colonial Insecurities in the Middle East", in Jutta Weldes et al. (ed.), *Cultures of Insecurity: States, Communities and the Production of Danger*, Minneapolis: University of Minneapolis, p. 152.

Putnam, Robert D. (1993): "The Prosperous Community: Social Capital and Public Life", *The American Prospect,* 13: 35-42.

Varshney, Ashutosh (2002): *Ethnic Conflict and Civic life: Hindus and Muslims in India*, New Haven: Yale University Press.

Varshney, Ashutosh (2001): "Ethnic Conflict and Civil society-India and Beyond", *World Politics,* 53 (3): 362-98.

Weldes, J. (1999): *Constructing National Interests: The United States and the Cuban Missile Crisis,* Minneapolis: University of Minnesota Press, p. 1.

3

Ethnic Violence in South Asia: A Media Perspective

Dharmesh Pushpa Dhawankar

Introduction

A sombre reality and disillusionment of our epoch, which emerged from the ashes of World War II, is that although there have been successes in the push toward development and modernisation, eradication of disease, and the spread of literacy, economic and political development programs have generated and stimulated, whether by collusion or in reaction, in good faith and poor anticipation, massive civil war and gruesome interracial and interethnic bloodshed. The same epoch has witnessed the rise in many countries of repressive authoritarianism in both military and democratic guises, equipped with Western weaponry, inflamed by popular slogans and fundamentalist doctrines, and assisted by a flagrant manipulation of the mass media, which have vastly expanded their reach.

The optimism of the sociologists and political scientists who naively foretold an "integrative revolution" and the inevitable decline of "primordial loyalties" such as kinship, caste, and ethnicity in third world countries, has by now waned and dimmed. The introduction of constitutions and democratic institutions enshrining human rights, universal franchise, the party system, elected legislatures, majority rule, and so on, has often resulted in strange malformation that are far removed from the goals of liberty, justice and tolerance that were the ideological supports of Western European and North American liberal democratic syntheses.

Something has gone awry in centre periphery relations

throughout the world and a manifestation of this is the occurrence of widespread ethnic conflict, accompanied in many instances by collective violence among who are not aliens but enemies intimately know. At different times certain ranges of phenomena, grouped under embracing labels such as "social class", "caste", "race", "gender inequality", "modernisation", "the colonial encounter" and so on, have become foci of intensified scholarly interest. Then these inquiries fade away, not only because of diminishing marginal return, but also because the phenomena themselves, as reflected on the screen of history, either salience or are transformed into other events, which are more revealingly grouped under new labels.

Other such label subsuming a range of phenomena with a family resemblance is "ethnicity". It is significant that the term ethnicity has come into vogue and found its way into Standard English dictionaries especially since the 1960s. Linguistics, national, religious, tribal, and racial divisions and identifications, and competitions and conflict based on them, are not of course, new phenomena, yet the recent salience of the term ethnicity reflects a new reality and a new usage reflects a change in that reality on a global scale in the latter half of the twentieth century, both in the industrialized first world and the "developing" third world.

It seems that the term ethnicity gained popularity in the social science literature of the 1970s, not only to describe certain manifestations in the third world, but also in reaction to the emergence of ethnic movements in the industrialized and affluent world, especially in the US, Canada, and Western Europe. And since the dismantling of the USSR and the other communist's regimes of Eastern Europe, the spates of internal conflicts that have exploded there are also viewed as ethnic.

The Reality Check

Ethnic conflict is a major reality of our time. This is confirmed, not simply by its ubiquity alone, but also by the cumulative increase in the frequency and intensity of its occurrence. According to a recent enumeration, some forty-eight countries are experiencing ethnic conflicts of one kind or other. The conflicts have occurred or been perpetuated between East Indian and Creoles in Guyana. Add to these upheavals that have become climatic in recent years among them the Sinhala-Tamil war in Sri Lanka, the Sikh-Hindu and Muslim-Hindu confrontations in Kashmir, and the turmoil engendered by the Ayodhya temple dispute, the Chackma-Muslim turmoil in Bangladesh, the actions of the Fijians against Indians in Fiji, the Pathan-Bihari clashes in Pakistan. Most of these conflicts have been involved violence, homicide, arson, and destruction of property.

Civilian riots have evoked action by security forces sometimes to quell them, sometimes in collusion with civilian aggressors, sometimes both in sequence. Events of this nature have happened in Sri Lanka, Bangladesh, Nepal, Burma, Pakistan and China. Mass killings of civilians by armed forces have been occurred in Indonesia, Pakistan and Sri Lanka. Some dissident ethnic groups have declared secessionist aims that threaten to break up extant polities, and these aims in turn have invited invasion of one country by another or armed interventions, such as by India in Sri Lanka. Moreover, ethnic conflict has also frequently caused massive displacements of people, many of them being deposited in refugee camps in neighbouring countries, as in south Asia.

The escalation of ethnic conflict has been considerably aided by gun-running and free trade in the technology of violence, which not only enable dissident groups to resist the armed forces of the state, but allow civilians to battle one another with lethal weapons. The classic definition of the state as the authority invested with the monopoly of force has become a sick joke. After so many successful liberations and resistance movements in many parts of the globe, the techniques of guerrilla resistance are now systematised and exportable knowledge.

Furthermore, easy access to the technology of warfare by groups in countries that are otherwise deemed low in literacy and in economic development we have seen what Afghan resistance can do with American guns is parallel by another kind of international fraternization among resistance group who have little in common save their resistance to the status quo in their own countries, and who exchange knowledge of guerrilla tactics and the art of resistance. Militant groups in Sri Lanka, Pakistan and India have international networks of collaboration, not unlike perhaps more solidarity than the diplomatic channels that exist between mutually wary sovereign countries and the great powers. There are, of course, global politico-economic processes at work, which cause disparities between countries, structures metropolitan centres and peripheries into configurations of uneven development, and stimulate neighbours of unequal power and status to confront one another.

The Forward Trend

Another development, not unknown in the past in the form of mercenaries for hire, but today reaching a sinister significance is the "privatization of war" that is, the ability of governments with extraterritorial geopolitical aims to fight their foreign wars, not by committing their own professional armies, but by farming out contracts for subversive military and political action to private professional groups willing to be hired or capable of being mobilized. The employment of ex-SAS veterans by the Sri Lanka government to help in the war against the Tamil militants is one of many such examples. Professionalized killing is no longer the monopoly of state armies and police force.

The internationalization of the technology of destruction, evidenced in terrorism and counterterrorism, has shown a face of free market capitalism and long distance trade in action unsuspected by Adam Smith and by Immanuel Wallenstein. The ubiquity, the increased frequency and intensity of ethnic conflicts, served by modern technology of destruction and communication, and published by the mass media, makes such conflict a special reality of the late twentieth century.

The Victorian perspectives in the main persisted perhaps until World War II, but there has been a recent change in paradigms, positing common world historical processes that hold centres and peripheries in one dialectical and interlocked field. These are specific developments that have contributed to a shift in historical consciousness that views our present world as a global village. The revolution in the media, instant transmission of visual images and auditory messages, linking metropolitan centres and distant places, and wide coverage of events, so that news broadcasts (whether NDTV, CNN, Star) present diverse events occurring at diverse places as a single synchronic and simultaneously occurring reality.

These communication processes bind us in a synchronicity of fellow witnesses of world events. We come to feel that the worldwide incidents of ethnic conflict are of the same order and are mutually implicated strife in killings of civilians in Sri Lanka, riots against Sikhs in Delhi, et al. These all belong to a contemporary world suffused by violence. The internationalization of violence and the simultaneity of its occurrences viewed on our TV screens make us all spectators and vicarious participants responding with our respective sympathies and our prejudices.

Has the Ethnicity Gone Political?

A major issue today is the transition from the politics of the nation-state to the politics of ethnic pluralism. It is useful at this juncture to take as our point of reference Benedict Anderson's *Imagined Communities: Reflections on the Origins and Spread of Nationalism*, in order to recognize its contribution and also to go beyond it to take account of newer developments. The versions of the 'dependency' theory of world capitalism variously proposed by Wallenstein, Andre Frank, Paul Baran, Samir Amin and Claude Merllassoux all in the end, post a monolithic global historical process.

However, they leave out of account the parallel process by which the world was fragmented, differentiated and carved up as 'nation-states'. Anderson shows how easily the modular concepts of the 'nation-state' and 'nationalism' were pirated by third world colonial and postcolonial elites. Under colonialism, the 'historical consciousness' of nineteenth century Asia was transmitted to an imbibed by local elites subjected to the textbook learning propagated by colonial schools. We have recently discovered with shocked surprise that the politics of ethnicity is not only a main preoccupation of the newly independent countries of the third world, but has emerged with a vengeance in what used to be called the second world. There appear to be two interrelated developments in this part of the world, which are characterized by both positive and negative features.

The collapse of totalitarian political systems has opened space for their replacement of democratic regimes. While one can see the obvious virtues of democracy and representative government compared with totalitarian Communist party rule, yet the fact is that the alleged changeover to electoral democratic methods in Asia has actually unleashed and/or intensified ethnic conflicts, and ethnic majorities have used their electoral dominance to inferiorize and assault or displace their minorities. Simultaneously, we are also witnessing a shake-up in the confidence and compliancy of Western Asia, which has recently shown disturbing evidence of chauvinism, racism, discrimination, identity panic, and a thrust towards national separateness, accompanied by attacks on foreign workers and migrants, as economies worsen and unemployment rises.

Our complacent assumptions of civilizational progress have been shaken. Lessons swept under the carpet and conveniently forgotten are resurfacing, that democratic politics and free-market capitalism have their prerequisites, their socio-cultural conventions and contexts have their own constructed myths, that they are not immune from their own internally generated contradictions and tensions, that they are not pills that surely and automatically cure illness.

There is another recipe for progress that is currently being tested, the idea of the nation-state as the desirable basis for structuring political life, a previously hegemonic concept that is currently undermined by regional and sub national resistances. Now, since the secular nation-state has been advocated by many Western theorists and third world intellectuals and political leaders as the bedrock on which modernization and economic development can be raised, it is extremely relevant to bear in mind two earnings.

First, historically, the Asian nations-state was achieved as the end result of very special developments, including social upheavals, internal strains, revolutions, and divisive wars between states.

Second, there is the possibility of a fundamental fallacy being perpetrated when an attempt is made to impose a historical construction such as the nation-state, achieved on distinctive soil, on a dependent world, as it its realization were a necessary stage in Universal History. This supposition, derived from Enlightenment assumptions, perhaps has near 'hegemonic' status in global affair.

What happens and indeed how do we perceive and represent and interpret what happens in many parts of the world where the events that led to the realization of the secular nation-state have not taken place, take place unevenly, or are actively resisted as harmful (as for example by Shiite fundamentalism in India by Hindu nationalists). It is now time to shift the language of 'obstacle' to 'development' to the language of active subaltern 'resistance' to it. In trying to sort out these issues, it is relevant to consider that the other side of the Asian model of the secular nation-state is its aggressive nationalism and imperialist expansion and penetration into what became its aggressive nationalism and imperialist expansion and penetration into what became its colonial dependencies.

Liberal democracies in Western Europe and the US frequently imposed authoritarian rule abroad, the exploitation of native labour and resources, and the inferiorization, if not erosion, of the cultures of the colonized. This inferiorization and threat of cultural extinction is large part lies behind the rise of Islamic fundamentalism, Buddhist 'nationalism', Hindu nationalism, and other such reactions, and their retaliatory attitude to western economic affluence and domination, political supremacy, alleged consumerist values, celebration of sexual eroticism, erosion of family durability, alleged 'privatization' of religion and separation of religion from affairs of state, and so on.

Need for an Assertive Dialogue

In a dislocating and sometimes disconcerting manner this hopeful, expansive phase of nation building has been put to the test and even reversed in the third phase, from the 1960s onward, by the eruption of ethnic conflicts. The divisiveness has revolved around issues of language, race religion, and territory. Accordingly, there has been a shift again in slogans and concepts. Ethnic groups and ethnic conflicts are the salient labels for talking about these events. The terms plural society, devolution of powers, traditional homelands, and self determination old words given new force and urgency have begun to frame the political debate and academic analyses.

The central political authority, the state, which in the previous phase of nation building and economic growth was designated as the prime actor and central intelligence in initiating, directing, and controlling the country's future and historical trajectory is now, after years of escalating ethnic divisiveness and pluralistic awareness, counselled to be a 'referee', adjudicating differences and enabling regional cultures and societies to attain their 'authentic' identities and interests.

The present phase characterizes by the politicization of

ethnicity, there are two salient features. The ethnic groups demand and bargain for collective entitlements, and it is usually a majority groups that demands affirmative action on its behalf to put to right an alleged historical injustice, thereby once giving new content to affirmative action, which is usually undertaken on behalf of depressed minorities and under classes. What I relate to the generation of regional or sub national reactions and resistances to what is seen as an over centralized and hegemonic state and their drive to achieve their own regional and local socio-political formations.

Now, let me enumerate four issues that have posed problems with regard to nation state making and 'modernization' in newly independent and so-called developing third world countries, and that have increasingly generated the politics of ethnics. They are the four rocks on which the nation-state project has foundered. The question of what the language or languages of education and administration ought to be, and of whether to replace English with swabhasha (indigenous languages), is a postcolonial problems that has taxed countries with plural languages, such as India, Pakistan, Burma, Sri Lanka, Malaysia, which all have their own written languages and literatures.

The ramifications of this language issue are many. Closely related is the 'modernization' program that has entailed the launching of ambitious literacy and educational programs. The result has been an explosion of literacy in the context of a population explosion, and the creation of large numbers of educated or semi-educated youths seeking employment in economies slow in growth and unable to accommodate them. It is this category of unemployed youth at urban sites that has everywhere been the most visible and activist participation in ethnic movements and violence.

A major divisive and contentious issue, generated by economic development and modernization in countries of low income and high population density and rural underemployment, is large scale population movements and migrations that cause dramatic speedy changes in the demographic ratios of people in a region who perceive themselves as different on the basis of ethnic origins, region, length of residence, and so on.

The second proposition is especially true when migration and the collision of groups produces 'competition for control over or access to economic wealth, political power and social status', and when there is a strong concept of 'territorial ethnicity' the notion that certain ethnic groups are rooted in space as bhumiputra (sons of the soil), especially among the indigenous fold of the region into which migrants are coming and when migration changes the demographic balance and the mix of ethnic groups within a given space. Migrants belonging to a particular ethnic group may move in from the periphery to work in subordinate positions to the ethnic group or nationality predominating in the core region. This situation results in a 'dual labour market' and applies to depressed minorities victimized by discrimination.

Discussion

The situation can become particularly acrimonious and contentious, especially in postcolonial and post independence times, when power shifts to and is exercised by the most numerous, usually the local 'sons of the soil; who then wish to displace these successful so-called aliens and newcomers. Frequently, this thrust coincides with the indigenous or local population producing its own educated youth, who aspire to move into occupations held and enterprise managed by the migrants. Such moves to displace people in favoured positions are particularly acute when the avenues of employment in the modern sector are not expanding fast enough to incorporate the number of entrants among the locals into the ranks of the middle class.

When such bottlenecks occur, the successful migrant are viewed as obstacles to the social mobility and well being of the indigenous majority. These known examples in illustrations, in northern India, in Assam and Tripura and elsewhere, the collisions between the local hill tribes and the incoming West Bengali Hindu and Bangladeshi Muslims migrants, in Pakistan, the animus against the Muhajir who migrated to Sind after partition and became prominent in Karachi face with similar displacement by 'indigenous' populations.

The issue pertaining to the degree of viability of secularism as specified in western nation state philosophy in civilizational contexts of the sort prevailing in many parts of the world in the middle east, in South and Southeast Asia, and elsewhere, where many persons reject the relegation of religion to the private domain and are earnestly committed to the idea that religious values and beliefs must necessarily inform politics and economic activities. The vexed issue is how to implement this world-view in a context where multiple religions with distinctive practices, and whose followers differ in their numbers, exist together in the same political arena.

It has been asserted that India has been the home of a concept of secularism, different from the western one, in which the state, rather than excluding religion from politics, is exhorted to be even-handed in its dealings with multiple coexisting religions that give direction to the lives of their adherents. Mahatma Gandhi, who once said that those who want to separate religion from politics understand neither, is held to be the quintessential proponent of the distilled wisdom of India that religions must and can coexist in a spirit of tolerance and mutual respect within the same polity.

Precedents for this were allegedly set by two celebrated Indian rules, Ashoka, who ruled over most of India, except the deep south, in the third century B.C., and whose 'righteous rule' was influenced by Buddhist values of tolerance and non-violence and the sixteenth century Mogul ruler Babur, who ecumenically reached out to Hinduism. These are indeed glorious precedents. But recent developments in India in the form of Sikh fundamentalism and the cry for Khalistan, and in the form of Hindu nationalism, propagated and propagandized with great effect by such organizations as the RSS, the BJP, the VHP, and the generated the Ayodhya temple dispute are putting India's capacity to negotiate a viable relation between a unified polity and sectarian religio-politics severely to the test.

There are similar developments in neighbouring countries, including Sri Lanka. The conundrum that faces many of us South Asians is this, while we all should make the effort to comprehend and appreciate the reasons for the rejection of Western secularism by certain religious communities, we also have to face up to the question of what policy to put in its place in an arena where multiple religious communities with divergent political agendas contest one another and make claims that threaten to engender discrimination and inequality among citizens who in principle must enjoy the same civil rights and should peacefully co-exist.

References

Ali, Ameer (1981): "The Racial Riots in Sri Lanka: 'A Reappraisal of Their Causes", *South Asia,* 4: 1-20.

Amnesty International (1987): "India: Allegations of Extrajudicial Killings by the Provincial Armed Constabulary in and around Meerut", 22-23 May, 1987. London amnesty International.

Anderson, Benedict (1985): *Imagined Communities: Reflection on the Origins and Spread of Nationalism,* London: Verso and NLB.

Basu, Amrita (1995): "Why Local Riots Are Not Simply Local: Collective Violence and the State of Bijnor, India", *Theory and Society,* 24: 57.

Bell, Daniel (1975): "Ethnicity and Social change", In Nathan Glazer and Daniel P. Moynihan (ed.), *Ethnicity: Theory and Experience*, Cambridge, Mass: Harvard University Press.

Buford, Bill (1992): *Among the Thugs: The Experience and the Seduction of Crowd Violence*, New York: Norton Press.

Canetti, Elias (1994): *Crowds and Power*, New York: Farrrar, Straus and Giroux.

Chakravarty, Uma, and Nandita Haksar (1987): *The Delhi Riots: Three Days in the Life of a Nation*, New Delhi: Lancer

International.

Chatterjee, Partha (1986): *Nationalist Thought and the Colonial World: A Derivative Discourse*, London: Zed Books for the United Nations University.

Connor, Walker (1973): "Nation Building or Nation Destroying: World Politics", *Journal of international Affairs,* 1.

Das, Veena (1990): *Mirrors of Violence: Communities, Riots and Survivors in South Asia.* Delhi: Oxford University Press.

Engineer, Asghar Ali (1993): 'Secular Forces Must Unite', *Frontline* (Interview), February 12.

Furnivall, J.S. (1956): *Colonial Policy and Practice,* 2nd ed., New York: New York University Press.

Golwalkar, M.S. (1990): *We or Our Nation Defined*, Nagpur: M.N. Kale Lecture.

Hensman, Rohini (1993): "Refugees: Their Potential as a Force for Reconciliation and Peace", *Pravada*, 1 (3).

Jeffrey, Robin (1986): *What's Happening to India? Punjab, Ethnic Conflict; Mrs. Gandhi's Death and the Test of Federalism*, London: Macmillan.

Kymlicka, Will (1991): *Liberalism, Community and Culture*, Oxford: Oxford University Press.

Rudolph, Lloyd I. (1993): *The Media and Cultural Politics in India; Votes, Alliance Politics and Minority Governments in the Ninth and Tenth General Elections,* Westview Press.

Taylor, Edwards B. (1973): *Primitive Culture: Researches into the Development of Mythology, Philosophy, Religion, Language, Art and Custom,* Vol. 1, London: John Murrey.

Weiner, Myron (1978): *Sons of the Soil: Migration and Ethnic Conflict in India,* Princeton, N.J.: Princeton University Press.

Yasmeen, Samina (1994): "Democracy in Pakistan: the Third Dismissal", *Asian Survey* 34 (6): 191-199.

4

Food Security in South Asia

S.S. Kalamkar

Introduction

Food security continues to be one of the important policy objectives of most of the developing countries in Asia, Africa and Latin America. At the global level, food production has exceeded the population growth rate. Nevertheless, despite this, food insecurity persists in many countries (Bhalla, 2004).

On a global level, food security for all requires that the supply of food be adequate to meet the total demand for food. While this is a necessary condition for the achievement of food security, it is by no means sufficient. Although currently enough food is supplied globally, as per *the State of Food Insecurity in the World 2008*, FAO estimates that 848 million people suffered from chronic hunger worldwide in 2003-2005 (representing an increase of 6 million from the nearly 842 million in 1990-92), in which 98.14 percent people (832.2 million) were in developing countries.

Fact that is more shocking emerged from recent provisional estimates of FAO that the number of chronically hungry people in 2007 were 923 million, increased by 75 million over 2003-05, with much of the increase attributed to high food prices. Given the continued and drastic price rises in staple cereals and oil crops well into the first quarter of 2008, the number of people suffering from chronic hunger is likely to have increased further. At 923 million people, the number of undernourished in 2007 was more than 80 million higher than in 1990-92, the base period for the World Food Summit (WFS) hunger reduction target.

This makes the task of bringing the number of

undernourished to 420 million by 2015 more difficult, especially in an environment of high food prices and uncertain global economic prospects. At the regional level, the largest increases in the number of undernourished people in 2007 occurred in Asia and the Pacific and in sub-Saharan Africa, the two regions that together accounted for 750 million (89 percent) of the hungry people in the world in 2003-05. FAO estimates that rising prices have plunged an additional 41 million people in Asia and the Pacific and 24 million in sub-Saharan Africa into hunger.

Food security in developing countries is a serious, multi-dimensional problem, distinctively in South Asians Countries. The *Agriculture Development Report, 2008* ranked South Asia as the second most undernourished, malnourished and food insecure region in the world. Achieving food security for its inhabitants remains a major challenge for South Asia (Babu, 2005). Although South Asian countries made impressive progress in foodgrains production during 1960s, 1970s and 1980s, the dynamism in the agricultural sector has however lost recently. Agriculture in the South Asian Region is caught in a low equilibrium trap with low productivity of staples, supply shortfalls, high prices, low returns to farmers and area diversification, results a threat to food security.

Productivity of major foodgrains crops has slowed and even declined for some crops (Kumar et al., 2008) and food production failing to keep pace with population growth. As result, South Asian countries are now finding it difficult to meet their populations' most basic food and nutritional needs and remain vulnerable to food security (Rasul and Schild, 2009). With this background, an attempt is made in this paper to study the extent of food insecurity in South Asian Countries.

The study is based on secondary data collected from the different sources. The data on various parameters of South Asian Countries is collected from the report *Human Development in South Asia* (MHHDC, 2009 and 2010), published reports and research papers and related websites.

Besides tabular analysis, annual growth rates were calculated to indicate an increase or decrease in various parameters.

Food Security: the Concept

As defined by World Food Summit (1996), *"Food security [is] a situation that exists when all people, at all times, have physical, social and economic access to sufficient, safe and nutritious food that meets their dietary needs and food preferences for an active and healthy life.* The concept of food security has evolved during last three decades to include not only food availability, but also economic access to food and biological absorption of food in the body. In fact adequate per capita availability of food is a function of the balance between food production on the one hand, and growth in population and purchasing power, on the other (Swaminathan, 2009).

Thus, food security and food insecurity are the terms used to describe whether or not all people at all time have physical and economic access to sufficient, safe and nutritious food for a healthy and active life. Thus, food security is multi dimensional and its major components are: (a) availability of food, (b) access to food, (c) quality and nutritional adequacy of food, and (d) utilization of food. Food in-security, can therefore, be a failure of one of these factors or some combination of these factors.

Food security is an essential foundation for meeting various Millennium Development Goals (MDGs) related to hunger, child mortality, mental health, disease, gender equality and primary education (Gill et al., 2003). Availability of sufficient food at the country or local level does not mean that all people within that geographical unit are food secure. Food security also implies that food intake must be adequate in both qualitative and quantitative terms. However, the issue of quality of diet has received little attention in the food security debate at all levels (Bose and Dey, 2005).

World Food Security

Importance of food management and maintaining food

security the world over has been abundantly recognized since long. Food security theme has been a consistent raised in specific contexts in a number of world conference convened by the United Nations in the 1990s (George, 2006). The necessity of having the food security at global level was first recognized at the 1974 World Food Conference where Food and Agriculture Organisation (FAO) was called upon to establish Committee on review and to recommend any action necessary in case of problems in particular areas.

The World Food Summit was called in response to the continued existence of widespread under-nutrition and growing concern about the capacity of agriculture to meet future food needs. In 1974, governments attending the World Food Conference had proclaimed that, "*every man, woman and child has the inalienable right to be free from hunger and malnutrition in order to develop their physical and mental faculties*". Subsequently, an international undertaking on World Food Security was set up in which Member Nations committed themselves to foods stocks at levels necessary to ensure continuity of supplies and to meet emergency needs in the event of crop failure or natural disaster.

Monitoring of implementation of the Undertaking has since then become the responsibility of the Committee on World Food Security. World Food Summit (1996) had set a target of reducing the number of hungry and food insecure to half (842 million in the base period of 1990-92) by 2015. Also, the first goal of UN Millennium Development Goals (MDGs) was to reduce the proportion of hungry (from a base level of around 20 percent) to half by 2015 (Acharya, 2009). The estimates shows that during the period of last three and half decades, there is reduction in population chronic undernourished in the developing countries (Table 4.1), by 17 percent points in South Asia, while the highest declined is noticed in East and South East Asia (32 percent points) in 2003-05 over 1969-71. Though the world has been changing remarkable over the past 25 years, food security still remains an unfulfilled dream for more than 800 million people living in the developing countries.

Table 4.1: Population Undernourished in the Developing Countries (percent to total population)

Region	1969-71	1979-81	1990-92	1996-98	2000-02	2003-05
Sub-Saharan Africa	34	37	35	34	33	30
Near East and North Africa	25	9	8	10	10	08
East and South East Asia	43	29	17	13	13	11
South Asia	38	38	26	23	22	21
Latin America and the Caribbean	19	13	13	11	10	8
All Developing Regions	37	29	20	18	17	16

Sources: FAO (2004 and 2008).

The global food security is in question today with ever-increasing food prices resulting from adverse climatic effects on agricultural production, rise in oil prices, increasing use of grains for bio-fuels, and almost 50 percent reduction in public spending on agricultural sector over the last three decades. The environmental sustainability has also become more elusive due to rapid industrial and population growth, urbanization and with lack of public realization about the sheer effects of environmental pollution. The food security has become increasingly important globally as well as at domestic front as the global food grains production is not keeping pace with increasing population (Singh and Mishra, 2008). From 1984 to 1994, annual growth in world food grain production was only one percent.

In fact, the global per capita availability of foodgrains has fallen substantially during the last decade of reforms, and the maximum decline has taken place during the first five years. The per capita annual production of cereals in the world increased from 271 kg. during 1961-65 to 295 kg. during 1966-70, which were the initial years of the green revolution (see, Kalamkar, 2009). The uptrend in production continued for about next two decades and per capita cereal production peaked by the mid-1980s at a level of 334 kg. per person per year, which was the highest one till date.

The per capita production of cereals declined to 310 kg. in the first five years of the last decade (2001-05). Though there is some improvement in per capita availability of cereals during the years 2003-07, this increase has not been available for use as food and feed, due to diversion of foodgrains for production of bio-fuel in United States and other countries. When total production is netted out for the corn used for bio-fuel, then per capita production falls to 307 kg., the lowest in any five-year period after 1966-70. This shows that the shortage of staple food has been building up over several years and it became quite large in the recent years (Chand, 2008).

Cereals constitute a major of diet of most of the developing countries like India. Cereal grain production kept pace with rising population in some nations, such as India. Many other countries turned increasingly to imports to feed their people, because either domestic production was too low or income rose far faster than population escalating demand for future food and feed grains. Although developing countries will account for most of the increase in global demand for cereals, growth in their demand is not as rapid as it was earlier. Due to diversifying diet and slower population growth in developing countries, growth in cereal demand of developing world is projected to decline from 2.3 percent a year in 1974-97 to 1.3 percent in 1997-2020 (Table 4.2). However, the world demand for meat and poultry is forecast to rise by more than 55 and 40 percent respectively between 1997 and 2020 (for details, see Rosegrant, 2001).

Table 4.2: All Cereals Production, Demand and Trade Data: 1997 and 2020

Particular	1997			
	Yield	Production	Demand	Net Trade
Developed Countries	3377	853996	724980	104334
United States	5334	337829	244430	76764
EU15	5540	203900	172599	20042
Former Soviet Union	1448	124264	131287	-2905
Developing Countries	2316	1017055	1118002	-104334
Latin America	2531	124584	138070	-15262
Sub-Saharan Africa	948	69303	82505	-12374
West Asia/North Africa	1975	85453	128991	-45080
South Asia	1814	235477	238325	-3020
Southeast and East Asia	6395	502204	529448	-27955
World	2703	1871051	1842983	0

Contd...

Particular	2020			
	Yield	Production	Demand	Net Trade
Developed Countries	3976	1024392	822259	202133
United States	6469	424481	305192	119290
EU15	5970	212527	183038	29489
Former Soviet Union	1647	144480	136219	8261
Developing Countries	3065	1472835	1674967	-202133
Latin America	3654	207400	210907	-3507
Sub-Saharan Africa	1380	128528	155872	-27346
West Asia/North Africa	2574	122474	195585	-73111
South Asia	2460	331797	353298	-21501
Southeast and East Asia	8480	682597	758188	-75590
World	3383	2497227	2497227	0

Note: yield–kg./ha; for net trade, positive figures indicate exports, negative figures indicate imports.
Source: IFPRI (2001).

As per the population projection estimates of the United Nations, the World population will increase to 8.3 billion in 2025 and 10 billion in 2050 from the present level of about 6.9

billion in 2010. Such unprecedented growth in population necessitates food production to be almost doubled by 2050. However, the strategies biased toward urban and industrial growth, together with the neglect of the agricultural sector, have led to significant damage to prime agricultural land and there is little evidence that the process of converting land to urban uses poses a serious threat to future global food production (Rosegrant et al., 2001).

The studies suggest that global agricultural production could be maintained relative to expected baseline levels over the next 100 years with moderate climate change. However, regional effects would be varying widely, and some countries may experience reduced output even if they take measures to adapt. Availability of land for farming is on the decline, and water for agriculture and other uses is increasingly scarce. Thus, together these and other forces will challenge the capacity of the world's food production system (Rosegrant et al., 2001).

Food Security in South Asia

South Asia (SA), comprising Bangladesh, Bhutan, India, Nepal, Pakistan and Sri Lanka has high population pressurc on land and other natural resources to produce food and meet other development needs (Paroda and Kumar, 2000). South Asia houses 22 percent of the world population on 3 percent of the land area. Half of the land area of South Asia is under arable and permanent crops, while in the World as a whole only 11 percent of the total land area is under arable land and permanent crops. In spite of land constraints, South Asian countries have made remarkable progress in food production during the past three decades, transforming the region from a food-deficit to a food self-sufficient region (Kumar et al., 2008).

Food self-sufficiency policies are the main basis of sector policies in the region. During the past 30 years, the research and agricultural support services of the system have generated growth in food production in excess of population growth, and reduced the proportion of people living in poverty. However, increased food production has not been fully translated in terms of household and individual food security. Despite of the high growth in food consumption, the region has the highest concentration of the poor and undernourished (299 million in 2003, WDR 2008) and accounts for 40 percent of the world's hungry.

South Asia is home to well over one fifth of the world's population, making it both the most populous and most densely populated geographical region in the world. South Asia is ranked as the second most undernourished, malnourished and food insecure region in the world. FAO estimates that even in 2010, Asia would account for nearly half of the world's malnourished population, of which two-thirds will be from South Asia (Mittal and Sethi, 2009). Agriculture is the predominant sector of economies of all South Asian countries and poverty and hunger are the most serious problems faced by this region.

Despite rapid growth of some of the economies in the region, dependence on agriculture as principal occupation has seen very small decline and thus future growth of agricultural sector holds the key to livelihood security, eradication of poverty, reduction in hunger, inclusive growth and sustainable progress of the economy of each of the nations. In the recent past, the growth of agricultural output has slowed down in most of the South Asian countries and in other countries where growth is still reasonably high level, it is becoming increasingly difficult to maintain that. South Asia has witnessed high annual growth rates of GDP in the period 1993-2006, while the growth rate of agriculture has been slower in the corresponding period. The deteriorating agricultural sector is said to add to the pressures of food supply and worsen the food security scenario (Mittal and Sethi, 2009).

Major changes have taken place in the food production scenario of South Asia over the last few decades. India, the largest of the South Asian economies, is now largely self-sufficient in food grain and an emerging exporter. On the account of the increasing availability of a wide variety of food items and increase in per capita income and urbanisation, the demand for better quality and nutritious food is now greater than in the past. During the period 1995-97 to 2001-03, South Asian countries witnessed a high annual growth rate of consumption that is often attributed to a high population growth rate (Mittal and Sethi, 2009).

One of the unique features of South Asia's structural transformation is that while the share of agriculture in total output has declined significantly and consistently over time, the share of the manufacturing sector in output has not picked up as predicted by economic theory (Table 4.3). The services sector continues to claim a higher share in output and employment, the share of manufacturing sector in output and, particularly, employment is not satisfactory.

The food security indicators in SA countries reveal that per capita income of South Asian countries is very low, which is principal reason for high incidence of poverty in the region. The lowest per capita income was in Nepal and highest was in Maldives in 2006-07. There is high level of inequality in this region (shown for instance by the Gini coefficients of income), something that needs to be taken into account in analyzing the high growth rates of food consumption. The rich in these countries consume much more than the poor and are responsible too for the high consumption growth rates, something that cannot be explained merely through overall population growth rates (Mitra, 2010).

Share of rural population in total population varied between 66 percent in Pakistan and 86 percent in Nepal. Small holders dominate the farm scene in South Asia except Pakistan. Also per capita availability of land resources in South Asia declining rapidly, has put serious pressure on agriculture and food security (Rasul and Schild, 2009). Productivity of crops is very closely related to irrigation and fertilisers use. Per hectare use of fertiliser was as low as 11.83 kg. in Maldives and as high as 287.31 kg. in Sri Lanka. While Pakistan has highest irrigated area and Bhutan has lowest among South Asian countries.

Table 4.3: Trends in Sectoral Composition of GDP: South Asia and Selected East Asian countries

Country	Manufacturing as percent of GDP			Agriculture as percent of GDP		
	1980	1990	2007	1980	1990	2007
Bangladesh	14	13	18	32	30	19
India	17	17	16	36	31	18
Nepal	4	6	8	62	51	34
Pakistan	16	17	19	30	26	21
Sri Lanka	18	15	19	28	26	12
South Asia	16	16	17	35	31	18
Malaysia	22	24	28	23	15	10
Indonesia	13	21	27	24	19	14
China	40	33	32	---	27	11
Korea, Rep.	24	27	28	16	9	3

Contd...

Country	Services as percent of GDP		
	1980	1990	2007
Bangladesh	48	48	52
India	40	41	52
Nepal	26	34	49
Pakistan	46	49	53
Sri Lanka	43	48	58
South Asia	41	43	53
Malaysia	36	43	42
Indonesia	34	42	39
China	21	31	40
Korea, Rep.	47	50	58

Sources: World Bank, 2008 and 2009, as quoted in MHHDC, 2010.

However, most disturbingly, the human development index (HDI) of this region shows that all the countries of this region have performed poorly in terms of HDI and HDI rank. Except Maldives and Sri Lanka, all other are ranked far above 100 in the latest HDI ranking. It is also a matter of great concern that according to the latest poverty estimates by the World Bank, using the US$ 1.25 a day poverty line, the total number of poor in South Asia has not come down since 1981 and in 2005 there were 596 million people below poverty line (World Bank, 2009).

The state of food insecurity in South Asia compared to East Asia during pre- and post-agricultural trade reforms is presented in Table 4.4.

Table 4.4: Trends in Incidence and Number of People Undernourished, 1990-2005

Country	Number of people undernourished (million)		
	1990-92	1995-97	2003-05
India	206.6	199.9	230.5
Pakistan	25.7	23.7	35.0
Bangladesh	41.6	51.4	40.1
Nepal	4.0	5.3	4.0
Sri Lanka	4.6	4.4	4.0
South Asia	282.5	284.8	313.6
East Asia	183.5	152.0	131.8

Contd...

Country	Percent undernourished in total population		
	1990-92	1995-97	2003-05
India	24	21	21
Pakistan	22	18	23
Bangladesh	36	40	27
Nepal	21	24	15
Sri Lanka	27	24	21
South Asia	26	22	21
East Asia	15	12	10

Source: FAO, 2009.

In 2003-05, South Asia accounted for 23 percent of the World's population but it contained 37 percent of the World's undernourished people. This share of undernourished people in South Asia increased by 0.7 percent points during the first half of the 1990s, while it accelerated by 2.7 percentage points between 1995-97 and 2003-05, an era of agricultural trade liberalization. In South Asia, during the post-reforms era, the proportion of undernourished people in the total population decreased by one percentage point compared to three percent points decline during pre-reform era.

The absolute number of hungry people increased by 28.8 million in the later period compared to an increase of 2.3 million in the former period. Among the countries, proportion of hungry in total population declined in Bangladesh, Nepal and Sri Lanka, however, it remained stagnant in India and increase in Pakistan during post-reform period. During post 1995 period, although progress was slow in East Asia, the overall number of hungry people declined in the region. This may be because of higher level of human development and effective institutional reforms in this region. In order to help poor and weaker sections of the society, large number of food security programmes are run by the South Asian Countries, while their magnitude, reaches and target population vary from country to country (Mittal and Sethi, 2009).

Food Consumption Pattern in South Asia

Food consumption expressed in kilocalories (kcal) per capita per day is a key variable used for measuring and evaluating the evolution of the global and regional food situation. A more appropriate term for this variable would be "*national average apparent food consumption*" since the data come from national Food Balance Sheets rather than from food consumption surveys (www.fao.org). Analysis of FAOSTAT data shows that dietary energy measured in kcals per capita per day has been steadily increasing on a worldwide basis; availability of calories per capita from the mid-1960s to the late 1990s increased globally by approximately 450 kcal per capita per day and by over 600 kcal per capita per day in developing countries (Table 4.5).

Table 4.5: Global and Regional Per Capita Food Consumption
(kcal per capita per day)

Region	1964-1966	1974-1976	1984-1986
World	2358	2435	2655
Developing countries	2054	2152	2450
Near East and North Africa	2290	2591	2953
Sub-Saharan Africa[a]	2058	2079	2057
Latin America & Caribbean	2393	2546	2689
East Asia	1957	2105	2559
South Asia	2017	1986	2205
Industrialized countries	2947	3065	3206
Transition countries	3222	3385	3379

Contd...

Region	1997-1999	2015	2030
World	2803	2940	3050
Developing countries	2681	2850	2980
Near East and North Africa	3006	3090	3170
Sub-Saharan Africa[a]	2195	2360	2540
Latin America & Caribbean	2824	2980	3140
East Asia	2921	3060	3190
South Asia	2403	2700	2900
Industrialized countries	3380	3440	3500
Transition countries	2906	3060	3180

Notes: [a] excludes South Africa.
Source: http://www.fao.org/DOCREP/005/AC911E/ac911e05.htm.

This change has not, however, been equal across regions. The per capita supply of calories has remained almost stagnant in sub-Saharan Africa and has recently fallen in the countries in economic transition. In contrast, the per capita supply of energy has risen dramatically in East Asia (by almost 1,000 kcal per capita per day, mainly in China) and in the Near East/North Africa region (by over 700 kcal per capita per day).

In case of South Asia, the per capita supply of energy has increased by 386 kcal in 1997-99 over 1964-1966. In short, it would appear that the world has made significant progress in raising food consumption per person.

Though, there is significant progress in raising food consumption per person in South Asia, the supply of foodgrains has increasingly lagged behind demand. In general terms past production growth rates for the main grain staples in South Asia have not kept up with consumption growth (Table 4.6). Except in Pakistan and Sri Lanka for rice, consumption growth of wheat and rice in South Asia has exceeded production growth by substantial margins. The gap between consumption and production growth is especially large for wheat in India and Pakistan, and rice in India.

Table 4.6: Annual Growth Rates of Production in Consumption of Rice and Wheat in South Asian Countries, 2000-08
(Average percent growth rate)

Crop	B	I	N	P	SL	SA	W
Production Growth							
Wheat	-8.8	0.3	2.1	0.3	0.0	1.6	1.8
Rice	2.0	1.5	0.2	1.8	2.2	0.2	1.0
Total	1.5	1.0	0.8	0.8	n.a.	1.0	1.5
Consumption Growth							
Wheat	0.0	1.9	2.2	1.2	2.0	2.2	1.3
Rice	2.6	2.6	0.4	-1.0	0.0	2.4	1.0
Total	2.3	2.3	0.9	1.0	1.6	2.3	1.2
Population Growth	1.6	1.7	2.3	2.6	0.9	1.9	1.1

B = Bangladesh; I = India; N = Nepal; P = Pakistan; SL = Sri Lanka; SA = South Asia; W = World.
Source: World Bank, 2010.

The average share of food consumption in total consumption is very high in South Asia, at close to 50 percent as compared to 17 percent in the United States. This high percentage makes South Asia very vulnerable to food price increases. High food prices undermine the food security of the poor. Food price inflation reduces purchasing power [1]; increases the number of people who go hungry (FAO, 2009). The share of food consumption expenditure in total household consumption expenditure in South Asian countries during 2002-07 presented in Table 4.7 shows that people living in rural areas spend a higher share of their income on food, and imported cheap food can harm the income of these people because a majority of them depend on the agricultural sector as either producer or worker (MHHDC, 2010).

Table 4.7: Food Consumption Expenditure in South Asia, 2002-07

Country	Share of Food Consumption Expenditure in Total Household Consumption Expenditure (2002-07)		
	National	Rural	Urban
India (2004)	49.5	54.0	41.6
Pakistan (2004)	47.6	53.6	39.6
Bangladesh (2005)	53.8	58.5	45.2
Nepal (2003)	59.0	62.9	39.1
Sri Lanka (2002)	44.5	46.2	35.9
Bhutan (2007)	39.2	44.8	32.5
Maldives (2003)	29.9	---	---

Source: FAO, 2009.

In South Asia, more than half of the population is employed in the agricultural sector and they spend about one-half of their income on food. Food security can significantly affect the learning and working ability of individuals and hence can create socio-economic problems. An increase in net food imports show more depend on the volatile world agricultural markets and loss of food self-sufficiency and sovereignty. For instance in 1998, India became the world's largest edible oil importer, just after achieving self-sufficiency in edible oil only five years age (PANAP, 2002). In fact, net food exports remained negative between 1990 and 2006 (FAO 2009) in all South Asian countries except India (MHHDC, 2010).

Long-term trend in consumption pattern at household level shows that per capita direct consumption of foodgrains has been declining and that of livestock products and fruits and vegetables has been going up for fairly long time in most of the SAARC countries. Despite this shift in dietary pattern, foodgrains are considered to be of paramount importance for household food and nutritional security. It is because of four reasons. One, cereals and pulses are staple foods and there is no perfect substitution between staple food and other foods. Two, due to inadequate intake of almost all foods, increased consumption of other foods, in most cases, fill dietary deficiency.

Three, foodgrains are the major and cheapest sources of energy and protein as compared to other foods and are thus vital for food and nutritional security of low-income masses. Four, increased production and consumption of livestock products resulting from rising per capita income require high growth in use of grain as feed for livestock. Because of these reasons, foodgrains continue to be main pillars of food security in South Asia and any slack in their production translates into persistent price shock and adverse impact on common people (Chand, 2010).

Food Demand Projections

The emerging production scenario, higher economic growth, population explosion and shift in dietary pattern will change the supply and demand prospects for food in the next few decades. South Asia will continue to exhibit a more diversified food basket with significantly higher levels of per capita consumption of milk, fruits, vegetables, meat, eggs and fish. Future increase in the production of food and non-food agricultural commodities has to be essentially achieved through increases in productivity, as possibilities of area and livestock population expansion are minimal. The increasing trends in consumption towards high value commodities will generate high growth in demand for pulses, livestock, fisheries and horticultural product. According to demand projects prepared by Paroda and Kumar (2000) under low income and high-income scenarios of South Asia, demand in the region would grow by 1 percent for cereals and by 1.7 percent for pulses, while demand for edible oils is projected to grow by more than 1.6 percent (Table 4.8). The demand for fruits and vegetables and livestock would rise by around 3 percent or more under high-income scenarios.

Table 4.8: Annual Growths in Domestic Demand for Food in South Asia: 1995-2030

S.N	Food	Annual growths in domestic demand for food (in percent)	
		Low income growth	High income growth
1.	Rice	1.2	1.2
2.	Wheat	1.3	1.2
3.	Maize	1.1	1.0
4.	Other coarse grains	0.9	0.8
5.	Total cereals	1.2	1.1
6.	Pulses	1.5	1.7
7.	Food grains	1.2	1.2
8.	Roots & tubers	1.6	1.9
9.	Edible oils	1.6	1.8
10.	Vegetables	2.3	2.9
11.	Fruits	2.3	2.9
12.	Sweeteners	1.5	1.6
13.	Milk	2.3	2.8
14.	Animal fat	2.3	2.8
15.	Meat	2.7	3.4
16.	Eggs	2.6	3.3
17.	Fish	2.5	3.2

Source: Paroda and Kumar (2000).

The commodity wise projected food demand in South Asian countries is presented in Table 4.9.

Table 4.9: Projected Demand for Food in South Asia
(million tonnes)

Commodity	Year	Bangladesh	India	Nepal
Rice	2015	25.81	98.11	3.79
	2025	28.96	106.58	4.55
Wheat	2015	3.28	74.75	1.47
	2025	3.71	79.86	1.74
Coarse cereals	2015	0.24	37.33	2.11
	2025	0.27	39.61	2.48
Pulses	2015	0.72	17.86	0.29
	2025	0.83	20.22	0.36
Foodgrains	2015	30.05	228.06	7.66
	2025	33.77	246.27	9.12
Edible oils	2015	0.79	11.19	0.18
	2025	0.92	12.58	0.22
Sweeteners	2015	1.47	32.73	0.20
	2025	1.68	36.40	0.24
Roots & tubers	2015	0.69	8.92	0.35
	2025	0.80	10.32	0.44
Vegetables	2015	2.24	102.18	2.45
	2025	2.67	125.48	3.16
Fruits	2015	1.89	68.66	0.77
	2025	2.29	84.75	1.00
Milk	2015	2.98	114.11	1.93
	2025	3.58	138.06	2.50
Meat	2015	0.59	7.54	0.42
	2025	0.73	9.62	0.56
Eggs	2015	0.25	2.84	0.04
	2025	0.31	3.62	0.06
Fish	2015	2.00	9.92	0.05
	2025	2.46	12.67	0.06

Contd...

Commodity	Year	Pakistan	Sri Lanka	South Asia
Rice	2015	3.75	2.17	133.63
	2025	4.55	2.35	146.99
Wheat	2015	30.13	1.00	110.62
	2025	35.97	1.09	122.37
Coarse cereals	2015	2.82	0.09	42.59
	2025	3.35	0.10	45.81
Pulses	2015	1.90	0.19	20.95
	2025	2.34	0.21	23.96
Foodgrains	2015	38.59	3.45	307.80
	2025	46.21	3.75	339.13
Edible oils	2015	2.46	0.08	14.70
	2025	3.04	0.09	16.84
Sweeteners	2015	6.38	0.66	41.44
	2025	7.80	0.74	46.86
Roots & tubers	2015	0.72	0.15	10.84
	2025	0.90	0.18	12.64
Vegetables	2015	8.91	1.04	116.82
	2025	11.48	1.29	144.07
Fruits	2015	10.28	1.17	82.76
	2025	13.47	1.40	102.90
Milk	2015	46.40	1.12	166.53
	2025	59.85	1.37	205.36
Meat	2015	3.09	0.15	11.79
	2025	4.03	0.19	15.13
Eggs	2015	0.57	0.07	3.77
	2025	0.74	0.09	4.81
Fish	2015	0.71	0.65	13.32
	2025	0.92	0.82	16.93

Source: Kumar et al. (2007).

It can be seen from the Table that by 2025, there would be a demand of about 339.13 million tonnes, comprising of about 147 million tonnes of rice, 122.4 million tonnes of wheat, 46 million tonnes of coarse cereals and 24 million tonnes of pulses. Besides foodgrains, South Asia demand would be about 17 million tonnes of edible oil, 144 million tonnes of vegetables and 103 million tonnes of fruits by 2025. Also 205 million tonnes of milk, 15 million tonnes of meat, 4.8 million tonnes of eggs and 16 million tonnes of fish would be required to meet its domestic demand in 2025. As mentioned earlier, shifting dietary pattern from foodgrains to non foodgrains, livestock and fisheries product would put pressure on foodgrains and oilcakes to meet feed demand for livestock (Kumar et al., 2010).

Degradation of Natural Resources and Food Security in South Asia

South Asian nations are faced with problems of rising population and declining farm land in the region. The rapid population growth and economic development are degrading environment through the uncontrolled growth of urbanization and industrialization, expansions and intensification of agriculture, and the destruction of natural habitats. The increasing resource degradation problems such as groundwater depletion, water logging, salinity, and land degradation would add to the challenges.

Mainly, food security of the poor will be at risk, as they would face severe resource constraints not only in accessing water, but also in investing in land and water management. Land, a non-renewable resource, is central to all primary production systems. Soil degradation is a serious problem in South Asian countries. About 83 million ha (mha) is affected by water erosion and 59 mha by wind erosion. In addition, loss of nutrients and organic matter content is a widespread problem throughout the regions, and affect about 26 mha of cropland area. Nutrient depletion is exacerbated by erosion.

The loss of cereal production in India is estimated at four million tonnes (mt) on moderately degraded land, and 11 mt on strongly regarded land, which together amount to 8 percent of India's total cereal production. This loss is equivalent to US$ 2.3 billion/year. Salinization is another problem, especially in the irrigated areas of the Indo-Gangetic Plains (Lal et al., 2011). In arid and semi-arid areas, the increased demand for water for agriculture and other uses is being met by excessive withdrawal of groundwater, leading to its depletion and quality deterioration.

Another ecological process that is supposed to have eroded the sustainability of agriculture is the soil fatigue caused by monoculture. The continuous cultivation of wheat, rice, sugarcane monoculture in some pockets of India over the last four decades, it is asserted, has resulted in a serious deficiency of micronutrients in the soil and has adversely affected its fertility. Among the other reasons, the virtual stagnation of these crops yields in recent years is attributed to the decline in soil fertility caused by this monoculture. Producing additional food with limited and degraded natural resources and providing economic access to food at the household level for ensuring food security for all will be the major challenge for these nations.

Climate Change and Food Security in South Asia

South Asia is home to over one fifth of the world's population and is known to be the most disaster prone region in the world. The high rates of population growth, and natural resource degradation, with continuing high rates of poverty and food insecurity make South Asia one of the most vulnerable regions to the impacts of climate change. Climate change is a major concern in SA because of alterations in temperature and precipitation, rise of sea level, melting of the Himalayan glaciers, and degradation of natural resources and the environment. According to the Fourth Assessment Report of the IPCC (2007), future projections of climate indicate that SA is very likely to warm during the twenty-first century.

Also, the fresh water availability is projected to decrease and coastal areas will be at greatest risk due to increased flooding at the sea and rivers. In some SA countries, a substantial decrease in crop yields from rain-fed agriculture could occur. Also, the irrigation sector will be strongly affect by climate change, as well as changes in the effectiveness of irrigation methods (Kundzewicz et al., 2007).

Additionally, dramatic changes in land use patterns in SA compound the problem of climate change. To cope with climate change more effectively in SA, it is necessary to identify integrated adaptation and mitigation options for a range of agro ecosystems so as to enable a favourable policy environment for the implementation of Regional Climate Change Adaptation Network. Majority of the poor people in SA are at risk because of the increase in frequency of extreme events, and especially the drought, floods and variability in climate. The projected impacts of climate change in South Asia will vary across sectors, locations and populations. Glacier melting is a cause of concern because of its impact on the availability of water, and thus agronomic productivity.

Temperature rise will negatively impact crop yields in tropical parts of South Asia where these crops are already being grown close to their temperature tolerance threshold. While direct impacts are associated with rise in temperatures, indirect impacts due to water availability and changing soil moisture status and pest and disease incidence are likely to be felt. Climate change may also impact the onset, distribution and amount of monsoons, as was the case in 2009 when monsoons failed in India. It is also feared that crop yields and agronomic production will be adversely affected, thereby exacerbating the food insecurity. The most significant impacts are likely to be borne by small-holder rain-fed farmers who constitute the majority of farmers in this region and possess low financial and technical capacity to adapt to climate variability and change (Lal et al., 2011).

Food Security in India: An Overview

Improving the food security is an issue of considerable importance for the developing countries like India where millions of people suffer from hunger and malnutrition. Due to deep rooted poverty, rapidly growing population, low agricultural productivity and resultant food and nutritional insecurity during early independence periods, country had to give high priority to make our population food secure which would in turn mean economic growth and reduce poverty.

Food security concerns can be traced back to the experience of the Bengal Famine in 1943 during British colonial rule, during which about 2 to 3 million people perished due to starvation. Since Independence, an initial rush to industrialize while ignoring agriculture, two successive droughts in the mid 1960s and dependence on food aid from the US exposed India's vulnerability to several shocks on the food security front (Nandakumar et al., 2010). That threat reached dangerous proportions in the mid-1960s, leading to the launching of the Green Revolution, achievement of food self-sufficiency, and subsequently, a growing stock of surplus food grains by the mid-1970s. There has been a significant change in the foodgrains scenario from a scarcity to a surplus situation because of the acceleration in the production of foodgrains in the 1980s. More significant, foodgrains stocks held by the government have increased beyond the requirements.

India's achievement since the 1960s and 1970s, of food security with respect to the production and distribution of staple crops, represents one of the most remarkable successes in agricultural development anywhere in the world. India has made impressive strides on the agricultural front during the last three decades. Foodgrains production more than doubled from 102 million tonnes in triennium ending 1973 to 218 million tonnes in 2009-10. Virtually all of the increase in the production resulted from yield gains rather than expansion of cultivated area under foodgrains which remained stagnant at around 124 million hectares.

The food policy and agricultural development strategy adopted by India to improve the food security situation paid rich dividends, and the ensuring improvements in food security can be accessed from several angles. The most significant change was the increase in the domestic output of foodgrains, particularly cereals. Statistics present a confusing picture of

India's progress on food security. Both per capita foodgrains consumption and total calorific intake have declined slightly in recent years among all levels of the population (Table 4.10).

Table 4.10: Annual Per Capita Consumption of Farm and Other Products in India (kg/p/y)

Group	1983	1987	1993	1999	2004-05
Cereal	140.29	138.67	123.02	120.67	72.9
Pulses	10.14	10.27	8.11	10.55	52.7
Milk	39.04	48.37	40.68	62.19	138.3
Edible oil	4.1	4.74	4.65	8.71	8.5
V&F	44.27	56.36	36.32	72.92	n.a.
MFE	4.69	5.1	4.94	6.19	n.a.
Sugar	9.69	9.99	8.92	12.05	n.a.
Other	31.21	29.81	24.21	32.96	n.a.

Notes: V&F=Vegetables and fruits; MFE= Meat, fish and eggs.
Source: Mittal (2006, 2008).

Though India was successful in achieving self-sufficiency by increasing its food production and also improved its capacity to cope with year-to-year fluctuations in food production, it could not solve the problem of chronic household food insecurity. With about 251 million below the poverty line, India accounts for about one-fifth of the world's poor (Kumar, 2005). Food availability and price stability, which are considered as a good measure of food security till 1970, were achieved through green revolution and Public Distribution System (PDS), however the chronic food security which is primary associated with poverty, still persisted in the country.

Despite self-sufficiency in food production, millions of people in India remain underfed and underemployed. Poverty is a major cause of food insecurity and sustainable progress in poverty eradication is critical to improve access to food. Over the years, the incidence of poverty, both rural and urban poverty has declined considerably. The percentage of population below poverty line declined from 52.3 percent in 1977-78 to 38.9 percent in 1987-88, and finally to 27.5 percent in 2004-05 (Table 4.11). However, the absolute number of poor or food-insecure people continues to be sizeable.

Table 4.11: Incidence of Poverty in India

Year	Poverty ratio (percent)			Number of poor (million)		
	Rural	Urban	Total	Rural	Urban	Total
1977-78	53.1	45.2	52.3	264.3	64.6	328.9
1983-84	45.7	40.8	44.5	252.0	70.9	322.9
1987-88	39.1	38.2	38.9	231.9	75.2	307.1
1993-94	37.3	32.4	36.0	244.0	76.3	320.3
1999-00	27.1	23.6	26.1	193.2	67.1	260.3
2004-05	28.3	25.7	27.5	220.9	80.8	301.7
As per estimates of Tendulkar Committee						
1993-94	50.1	31.8	45.3	327.7	74.9	403.0
1999-00	41.8	25.7	37.2	298.0	73.1	371.0

Note: Uniform recall period
Source: GoI, (2011a, Mid-Term Appraisal of Eleventh FYP).

Though physical access to food is achieved, economic access at micro-level lagged behind indicating food and nutritional insecurity. One quarter of the world's hungry people live in India. *The State of Food Insecurity in the World, 2008* released by the FAO estimated that India is home to more than 230 million undernourished people, which is 21 percent of the national population. India is home to the greatest population of severely malnourished children in the world. Four hundred million children suffer daily, which is greater problem than in Sub-Saharan Africa.

Accordingly, to the *National Family Health Survey of India*, 55 percent children living in rural areas suffer from malnutrition compared to 45 percent of children in urban areas. In addition to this, per capita per day availability of food grains in India is declined, which was 468.8 gms. in 1971 and decreased to 444 gms. in 2009. The per capita availability of foodgrains has fallen substantially during the last decade of reforms, and the maximum decline has taken place during the last five years.

The poor agricultural productivity and production, and low level of foodgrains output resulting from the low level introduction of agricultural/crop technologies; poor rural infrastructure; high vulnerability of crop production to natural disasters such as floods and droughts; and high rates of unemployment and poverty, are some of the reasons for the high degree of food insecurity (Kumar, 2003). Further, it is estimated that with a population of almost 1.2 billion today, and likely to be the most populous country on this planet by 2030 with 1.6 billion people, currently accounting for more than 17 percent of the global population and 456 million poor or 41.6 percent living on less than US$ 1.25 a day (Chen and Ravallion, 2008), ensuring food and nutrition security is a challenge for India (Nandakumar et al., 2010).

The Indian government is more serious on this issue and therefore, the National Development Council (NDC) of India in its 53rd meeting held on 29th May, 2007 adopted a resolution to launch a '*National Food Security Mission (NFSM)*' comprising rice, wheat and pulses to increase the production of rice by 10 million tons, wheat by 8 million tons and pulses by 2 million tons by the end of the Eleventh Plan (2011-12), which has been launched from 2007-08 (http://agricoop.nic.in).

Conclusions

Agriculture is the predominant sector of economies of all

South Asian countries and poverty and hunger are the most serious problems faced by this region. Agriculture in this region is caught in a low equilibrium trap with low productivity of staples, supply shortfalls, high prices, low returns to farmers and area diversification, results a threat to food security.

Despite rapid growth of some of the economies in the region, dependence on agriculture as principal occupation has seen very small decline and thus future growth of agricultural sector holds the key to livelihood security, eradication of poverty, reduction in hunger, inclusive growth and sustainable progress of the economy of each of the nations. In the recent past, the growth of agricultural output has slowed down in most of the South Asian countries and in other countries where growth is still reasonably high level, it is becoming increasingly difficult to maintain that. The environmental sustainability has also become more elusive due to rapid industrial and population growth, urbanization, and with the lack of public realization about the sheer effects of environmental pollution.

As India plays a pivotal role in the geopolitics of the South Asia, weakening agricultural performance during last one decade in a large number of Indian states is a major concern. The situation is being further complicated by decreasing per capita availability of arable land and climate change risks, which will have adverse impacts on food production, compounding the challenge of meeting food demand. India has solved the problem of food security, which is reflected in mounting buffer stocks, but there are millions of food insecure and undernourished people in India.

The latest high economic growth rates have failed to improve food security in India leaving the country facing a crisis in its rural economy. The per capita availability of foodgrains has fallen substantially during the last decade of reforms, and the maximum decline has taken place during the first five years of last decade. Therefore, suitable land use pattern needs to be implemented, based on the principle that each region would focus on crops best suited to agro-climatic characteristics, soil types and water resources.

End Note

1. In developing countries as a group, food consumption expenditure decreases by an average of about 0.75 percent for every 1 percent increase in the price of food (Regmi et al., 2001).

References

Acharya, Shabd S. (2009): "Food Security and Indian Agriculture: Polices, Production Performance and Marketing Environment", *Agricultural Economics Research Review*, 22 (1): 1-20.

Babu, Suresh (2005): "Food Security in South Asia", *South Asian Journal,* July-September, (http://www.southasianmedia.net).

Bose, M.L. and M.M. Dey (2005): "Food and Nutritional Security in Bangladesh: Going beyond Carbohydrates Counts", *Agricultural Economics Research Review,* 20 (2): 203-225.

Chand, Ramesh (2008): "The Global Food Crisis: Causes, Severity and Outlook, *Economic and Political Weekly,* 43 (26-27): 115-123.

Chand, Ramesh (2010): "SAARC: Agricultural Vision 2020", *Agricultural Economics Research Review*, 23 (2): 197-208.

Chen, Shaohau and Martin Ravallion (2008): 'The Developing World is poorer than we Thought, But No Less Successful in the Fight against poverty', *Policy Research Working Paper 4703,* Washington, D.C.: The World Bank Group.

FAO (2004, 2008): *The State of Food Insecurity in the World 2008*, Food and Agriculture Organization.

FAO (2009): FAOSTAT, *http://faostat.fao.org* (accessed October 2009).

FAO (2009): *The State of Agricultural Commodity Markets*, Food and Agriculture Organisation, Rome.

George, P.S. (2006): "Some Reflections on Food Security in India", in N.A. Mujumdar and Uma Kapila (eds.), *Indian Agriculture in the New Millennium-Changing Perceptions and Development Policy*, Academic Foundation, New Delhi, pp. 194-226.

Gill, G.J., J. Farrington, E. Anderson, C. Luttrell, T. Conway, N.C. Saxena, and R. Slater (2003): 'Food Security and the Millennium Development Goal on Hunger in Asia', *Working*

Paper 231, UK: Overseas Development Institute.

GOI (2011): *Agricultural Statistics at a Glance, 2010,* New Delhi: Directorate of Economics and Statistics, Ministry of Agriculture, Government of India.

GOI (2011a): Mid-term Appraisal of Eleventh Five Year Plan, Planning Commission, Government of India, New Delhi, (http://www.planningcommission.gov.in/plans/mta/11th_mta/pdf/MTA_comp11th.pdf).

IFPRI (2001): IMPACT Projection, June, (http://www.ifpri.org/pubs/books/gfp/gfp_apxd.pdf).

IPCC (2007): *Climate Change 2007: The Physical Science Basis, Summary for Policymakers,* Contribution of Working Group I to the Fourth Assessment Report of the Intergovernmental Panel on Climate Change.

Kalamkar, S.S. (2009): "Globalisation, Food Security and Sustainable Agriculture in India", *The Indian Economic Association,* 92nd IEA Conference Volume, Part II, pp. 90-113.

Kerr, John (1996): "Sustainable Development of Rain-fed Agriculture in India", *EPTD Discussion Paper No. 20,* Environment and Production Technology Division, Washington, D.C.: International Food Policy Research Institute.

Kumar, M. Dinesh (2003): "Food Security and Sustainable Agriculture in India: The Water Management Challenge" *Working Paper No. 60,* Colombo: International Water Management Institute.

Kumar, Praduman (2005): "Empowering the Small Farmers towards a Food Secure India", in *India's Agricultural Challenges: Reflections on Policy, Technology and other Issues*, CENTAD Publications, New Delhi, pp. 197-225.

Kumar, Praduman, Surabhi Mittal and Mahabub Hossain (2008): "Agricultural Growth Accounting and Total Factor Productivity in South Asia: A Review and Policy Implication", *Agricultural Economics Research Review,* 21 (2): 145-172.

Kundzewicz, Z.W., Mata, L.J., Arnell, N.W., Döll, P., Kabat, P., Jiménez, B., Miller, K.A., Oki, T., Sen, Z. and Shiklomanov, I.A. (2007), "Freshwater Resources and their Management", in Parry, M.L., Canziani, O.F., Palutikof, J.P., Van der Linden, P.J., Hanson, C.E. (eds.), *Climate Change 2007: Impacts, Adaptation and Vulnerability*, Contribution of Working Group II to the Fourth Assessment Report of the Intergovernmental Panel on Climate Change. Cambridge University Press, Cambridge, pp 173-210.

Lal, R.; M.V.K. Sivakumar; S.M.A. Faiz and A.H.M. Murtaflzur Rahman, and K.R. Islam (2011): *Climate Change and Food Secuiry in South Asia,* Springer Science+Business Media B.V.

Lumar, Praduman; Mruthyunjaya and Ramesh Chand (2010): "Food Security, Research Priorities and Resource Allocation in South Asia", *Agricultural Economics Research Review,* 23 (2): 209-226.

MHHDC (2009): *Human Development in South Asia 2008: Technology and Human Development in South Asia*, Karachi: Oxford University Press.

MHHDC (2010): *Human Development in South Asia 2009: Trade and Human Development,* Mahbub-ul-Haq Human Development Centre, Pakistan: Oxford University Press.

Miital, Surbhi and Deepti Sethi (2009): *Food Security in South Asia: Issues and Opportunities,* Working Paper No. 240, New Delhi: Indian Council for Research on International Economic Relations.

Mitra, Amit (2010): "Food Security in South Asia: A Synthesis Paper", available at, http://www.idrc.org/uploads/user-S/12858517301Food_Security_in_South_Asia.pdf).

Mittal, Surabhi (2006): *Structural Shift in Demand for Food: Projections for 2020,* New Delhi: Indian Council for Research on International Economic Relations.

Mittal, Surabhi (2008): "Demand and Supply Trends and Projections of Food in India", *Working Paper No. 209,* New Delhi: Indian Council for Research on International Economic Relations.

Nandakumar, T.; Ashok Gulati; Pravesh Sharma and Kavery Ganguly (2010): "Food and Nutrition Security Status in India: Opportunities for Investment Partnerships", Food for All Investment forum for Food Security in Asia and Pacific, ADB-FAO-IFAD.

PANAP (2002): *Impact of Agreement on Agriculture and Trade Liberalization on Food Security,* Penag: Pesticide Action Network Asia and the Pacific, Ganesh Printers.

Paroda, R.S. and Praduman Kumar (2000): "Food Production and Demand in South Asia", *Agricultural Economics Research Review,* 13 (1): 1-24.

Rasul, Golam and Andreas Schild (2009): "Food Security and Agricultural Sustainability in South Asia: A Mountain Perspective", in proceedings of *National Seminar on Food Security and Sustainability in India,* Amritsar: GAD Institute of Development Studies, pp. 32-41.

Regmi, A., M.S. Deepak, J.L. Seale Jr., and J. Bernstein (2001): "Cross Country Analysis of Food Consumption Patterns", in Regmi, A. (ed.), *Changing Structure of Global Food Consumption and Trade,* Washington DC: USDA ERS.

Rosergrant, Mark W., Michael S. Paisner, Siet Meijer and Julie Witcover (2001): *2020 Global Food Outlook-Trends, Alternatives, and Choices,* Washington, D.C.: International Food Policy Research Institute.

Singh, N.B. and J.P. Mishra (2008): "Towards Sustainable Foodgrains Production for National Food Security", *Agricultural Situation in India,* LXV (5): 267-280.

Swaminathan, M.S. (2009): "Strengthening Food Security Based on Home Grown Foods", *Indian Journal of Agricultural Economics,* 64 (2): 177-186.

UNDP (2007): *Human Development Report 2007-08*; *Fighting Climate Change: Human Solidarity in a Divided World,* New York: United Nations Development Programme.

World Bank (2008): *World Development Report,* Washington, D.C.: World Bank.

World Bank (2008, 2009): *World Development Indicators,* Washington, D.C.: World Bank.

World Bank (2010): 'Food Price Increase in South Asia: National Responses and Regional Dimensions', Agriculture and Rural Development Unit, Sustainable Development Department, *South Asia Region, The World Bank,* Washington, D.C. 20433.

World Food Summit (1996): *Rome Declaration on World Food Security,* (http:www.fao.org).

5

India's Innovation Systems: Changing Structure in the Post-TRIPS Regime

Girish Kumar R.

Innovation 'is not an end in itself but a means to productivity growth and higher living standards' (World Bank: 2007). India is emerging as a global hub of innovation-low cost as well as high value products and services (Kelkar, et al., 2006; Herstatt, 2008). India was ranked as the most preferred destination for locating Innovation centres. Many multinationals have set up R&D laboratories in India, for example, General Electric in the John. F. Welch Technology Centre in Bangalore, which in 2000 was their largest single R&D location outside the US. The ultimate reason for the relocation of multinationals to India and China is the high return on R&D investment, realized through the skilled R&D workers. Mr. Welch, the then chief operating officer (COO) of General Electric, commented, "India is a developing country, but it is a developed country as far as its intellectual infrastructure is concerned.

We get the highest intellectual capital per dollar here" (Tripathi, 2007). The Government of India has played a crucial role in transforming its national innovation systems to adapt to the current situation. The first major initiative was the Scientific Policy Resolution enacted in 1958 to support early and critical stages of industrialization. The next was the Indian Patents Act of 1970, which granted process patents. This Act was instrumental in making Indian firms to adapt reverse engineering. The Act was amended in 1999 following a WTO dispute and was subsequently made TRIPS-compliant in 2003 (Mani, 2004). In 2003, over 70 percent of the R&D

expenditure was from Central or state governments enterprises, which accounted for a mere 25 percent (GoI, 2006).

The trend is changing. The government ended the traditional R&D policy according to which subsidies and grants for R&D were the principal public support mechanism. National laboratories of the Council of Scientific and Industrial Research (CSIR) and other prestigious technical institutes were instructed to earn at least 30-50 percent of their R&D expenditures through commercialization of indigenously developed technologies and the generation and utilization of patents (Sikka, 1998).

The main focus of its reforms has been to link the science sector with the business sector and to provide incentives for innovation activities. The Science and Technology Policy 2003, recognizing the changing context of the scientific enterprise, and to meet present national needs in the new era of globalization announced a new S&T policy in 2003 which ambitiously set the goal of R&D spending at 2 percent of GDP by March, 2007.

The policy articulated eleven strategies for achieving the objective. The policy also had four new features: (i) it recognized that India has a large pool of scientists and engineers, but a low density; (ii) it acknowledged the necessity to manage the brain-drain from the country; (iii) it highlighted the trend of increasing patenting both at home and abroad; and (iv) it noted the need to monitor implementation of the policy. Despite the aggressive goal, the feasibility of implementing the policy has been questioned (Mani, 2004).

Since 1991 Indian firms have been exposed to market forces where innovation increasingly plays a key role in ensuring economic survival and achievement. India tried to make technological advancements by balancing the import of technology and indigenous R&D efforts (Fan, 2008). Global firms have used three options for sourcing innovations from India, namely: (i) locating centres of Innovations in India through fully owned subsidiaries; (ii) outsourcing innovations to Indian firms or research centres, and (iii) acquiring innovative entrepreneurial firms or start-ups (Bowonder et al.). Indian companies are also making acquisitions abroad to leverage the various possibilities of innovation that the global market offers.

Besides, the existence of an open society, a technology base, democracy, diversity, an environment that allows experimenting, a vibrant capital market, availability of young population necessary as human capital to fully reap demographic dividends, full and free competition in the private sector and opportunities for technological leapfrogging are all factors that make India a favourable destination for innovation.

Indian firms are still in the mindset of incremental innovation as compared with breakthrough Innovations. Firms with majority foreign ownership have greater Innovation Intensity than those with majority Indian ownership. Similarly, Innovation Intensity for MNCs is significantly higher than for non-MNCs (National Knowledge Commission, 2007). Innovation is confined to the pharmaceutical industry, which enabled India to become a global leader in the production of generic drugs and bring 'pharmacy of the poor'.

However, R&D in this sector is confined to the production of process patents than product patents. Like any other developing nation, it is difficult for Indian pharmaceutical companies to develop a new product given their resource constraint. Still it is amazing to note that the productivity of innovative outputs is very high compared to other nations. Similarly, the cost of Indian scientists engaged in R&D is comparatively very low, making it a favourable destination for pharmaceutical outsourcing.

However, the technological advancements are too quick in other industries like electronics, telecommunications and engineering rendering patents not too relevant, while the impact of patents is profoundly felt in the pharma industry [1] where the patent holder can effectively hold monopoly over a particular product for 8-10 years. During TRIPS compliance period many pharmaceutical companies like Dr. Reddy's, Cipla and Ranbaxy [2] (no longer Indian), among others, were forced to increase their R&D expenditure. But they are all small companies compared to MNCs, whose R&D expenditure alone constitute the total budget of Indian minnows. Presently, they develop new chemical entities (NCEs), which are later licensed to MNCs. [3] India made its impact in the area of ICT software as well (Fan). It is also emerging as a global player in the biotech industry.

In other industries, there are sporadic instances of new product developments. Interestingly, the only firms that named R&D as the most important area of focus were pharmaceuticals firms. Still it should be noted that India is facing a shortage of R&D manpower, for research needs qualitatively superior engineers and students. Despite having a large pool of low-cost scientists, university education cannot meet the standards of the industry. India is looking to attract expatriates from overseas into high-level elite positions in the hope of repeating Taiwan's success story in the electronics industry (Fan, 2008: 17).

The current status of India's patent policy, at least from the point of view of substantive law, is well-matured having stated that there are a lot of procedural issues that need to be ironed out. Procedural issues mainly relate to lack of awareness amongst various stakeholders, non-availability of qualified examiners at the patent offices, absence of any special court to deal with patent infringements and invalidation as we, as poor enforcement of existing laws by law enforcement agencies. However, on a positive note, India is presently contemplating a law on the lines of the Bayh-Dole Act of USA.

The Bayh-Dole Act permits universities and their faculty members to stake patent claims on discoveries they made through research funded by federal agencies instead of leaving the ownership of the intellectual property with the government. Such a law is expected to trigger R&D efforts and promote technology transfer by creating incentives for university researchers to consider the practical applications of their discoveries, and for universities to scout for potential companies to develop them. It is expected that the Indian Parliament will pass the Law by the end of 2008.

Intellectual Property (IP) Policy and Innovations

Procedural obstacles encountered in the process of patent application can be of the following types—lack of information regarding the prior art of invention (Orazco, 2007); non-availability of information relating to the technical features of the invention that prevents attorneys from drafting the specifications [4] and the shortage of trained manpower possessing the ability to grasp the technology and law. [5] Confusing definition of 'inventive step' as 'an invention that involves technical advance as compared to the existing knowledge or having economic significance or both' makes the invention not obvious to a person skilled in the art and hence delays the entire process. The provision of pre-grant opposition is sometimes misused by frivolous opposition, which delays patent applications (Bird and Cahoy, 2007).

Patent holders can authorize third parties to work on the invention by getting into voluntary arrangements. Such voluntary licenses can be one-sided and it can detrimentally affect access to affordable healthcare as many of these voluntary licenses can involve exorbitant costs to the generic companies (Mathew and Prasad, 2005). During the bird flu crisis of 2005, two Indian generic pharmaceutical companies, Cipla and Ranbaxy, approached Roche to work out a voluntary licensing deal, but it did not materialize (Rai, 2008). Furthermore, working out a voluntary license agreement between a generic manufacturer and a brand name company is a relatively rare occurrence and is often difficult to negotiate, with the brand name company placing limits on production capacity, demanding high royalty fees or imposing other similar restrictions (Basheer, 2007).

The Indian Patent and Trademark Office (PTO) in 1993-94

spent about US$ 330,000, whereas the United States PTO in the late 1980s spent approximately US$ 300 million a year (Gupta, 2005: 602). Additional funds, however, are gradually being devoted to the development of a stronger intellectual property administration infrastructure, but India is still spending far less on their patent infrastructure than such countries as the US. In addition to the smaller budget for intellectual property, there exists evidence to suggest that India's judiciary is, as a rule, not well versed in economic theory and often makes decisions that are hostile to good economic judgment.

Thus, when the negative aspects of patents, such as higher prices, are immediate and easy to see, and the upside of increased patent protection is more long-term and more obscure, it could affect judicial decision-making (Colin, 2007). Another possibility is the engagement of comparatively junior lawyers at the lower courts by NGOs against the MNCs at the courts. This sometimes enables the latter to get a favourable verdict even if they have a weak case.

Needless to say, a mismanaged patent portfolio can produce devastating results for the company. In India there is no coherent patent management strategy, which leads to breakdowns in both the acquisition and exploitation of a business's intellectual property assets. Following are the possible factors which act as a barrier to the failure of patent management to obtain critical patents or patents that may lead to lucrative situations; failure to identify potential relationships with partners and licensees; failure to realize the defensive value of a patent portfolio when challenged by another company on intellectual property issues; and failure to follow through on patent-related aspects of business, such as product development, branding and market expenditures (Osenga, 2007: 422).

Indian IP law in the post-WTO regime is becoming more stringent to make it TRIPS-compliant giving the patent owner a certain market monopoly for a period of time. Thus, patents erect barriers for transfer of technology from the North, which always relies on royalty payments and heavy license fees. This forced the closure of many small-scale industries that relied on reverse engineering, especially in the R&D-centric pharmaceutical sector. Earlier, patents were used as a government policy tool instead of an individual property right alienable within technology markets. The IP policy was to regulate access to innovations sensitive to the socio-economic concerns of India. Thus, the Indian Patent Act of 1970 provided for compulsory licensing of food, drugs, medicinal or chemical innovations, whereby any party can license the technology three years after the patent issuance.

Indian IP policy also rests on a philosophical foundation that extends to the domains of environment and healthcare. The 1970 Act prohibited the patenting of methods of agriculture and horticulture. Similarly, any processes for the medicinal, surgical, curative, prophylactic or other treatment of human beings to render them free of disease were put beyond the patents. This IP policy made India a pharmacy of the South with a large generic pharmaceutical industry having more than sixteen thousand firms. These firms were well suited to reverse-engineer pharmaceuticals developed and patented in foreign countries, and to design a new process for producing the same patented drug in large quantities.

Thus, the Patents Act, 1970, gave the Indian generic pharmaceutical industry great competitive advantages by allowing Indian firms to copy patented pharmaceutical products developed by foreign pharmaceutical companies by simply designing a new method to make the same patented drug. The Patents Act, 1970 was also instrumental in controlling drug prices in India and ensuring access to medicines to the bulk of its poor population. For example, ten tablets of 150 mg. dose Ranitidine Zantac costs ₹ 7.17 in India. In Pakistan it costs ₹ 122.16. In the United Kingdom, it costs ₹ 320.85. In the United States, these ten tablets cost ₹ 730.62. Additionally, the Act placed the burden of proof on the patentee to prove infringement (Colin, 2007). The Patents Act, 1970 was intended to stimulate the indigenous Indian pharmaceutical industry. Consequently, the multinationals were dislodged from their dominant position and became a junior partner in the industry.

Table 5.1: Major Indian Policy Initiatives to Encourage Innovations

Year	Policy Initiative
1958	Scientific Policy.
1970	Indian Patent Act.
1983	Technology Policy Statement.
1985	R&D Cess Act.
1988	Venture Capital Guidelines Announced.
1995	Technology Development Board Act.
1996	CSIR 2001: Vision and Strategy Announced Securities and Exchange Board of India (Venture Capital Funds) Regulations.
1999	The first amendment to the Indian Patent Act (IPA)1970: to put in place a mechanism for accepting product patent applications covering pharmaceutical and agriculture chemicals from January 1, 1995 and to provide exclusive marketing rights if certain conditions are fulfilled.
2000	New Millennium Indian Technology Leadership Announced.
2002	The second amendment to the IPA to bring it in conformity with all the relevant provisions included in the TRIPS Agreement, barring a solitary exception. This exception viz., introduction of product patents in the area of chemicals, pharmaceuticals.
2003	New S&T Policy Announced.
2005	The third amendment to the IPA extending product patents to chemicals, pharmaceuticals, agricultural chemicals and food.

Source: S. Mani (2007: 152).

Indian membership at the WTO saw changes in its IP

policy to make it TRIPS-compliant. [6] This also coincided with the Government's policy of liberalisation. This policy shift occurred in three stages. First, the economy was opened to foreign investment leading to a proliferation of joint ventures and foreign partnerships. Multinational companies began patenting locally. However, it did not add much value to Indian technological leadership.

During the second stage, the emphasis was placed on the development of home-grown technologies. Strategic focal changes to develop market-driven indigenous technologies originated in science and technology organisations, such as the CSIR and the Indian Institute of Science (IISc). The final stage required an efficient legal infrastructure to protect these technologies. Evidence shows that the policies have slowly begun to work. Annually, domestic patent applications filed in the Indian patent office have increased by 17 percent—2,247 patent applications were filed during the period 1970-1999. Foreign applications increased even more, with more than 6,500 patent applications filed during the same period. The number of patents granted in India increased from 22 in 1995 to 347 in 204 (Mani, 2007).

After four decades of inaction, Indian R&D began to change dramatically from the year 1991, when the country started its economic liberalisation process. This paved the way for a series of events that altered the country's R&D landscape forever. Till 1991, Indian R&D institutions enjoyed a high level of state patronage.

At that time R&D institutions consisted entirely of national laboratories and a few high-profile educational institutions like the Indian Institute of Science (IISc) and the Indian Institute of Technologies (IITs). The Government was determined to make growth 'inclusive'. The contribution made by the private sector to R&D was negligible. Although some MNCs had started taking interest, only Texas Instruments and Motorola made any sort of commitments in India. But things are changing now with India fast turning into a research hub.

IP and innovation policy have been designed in such a way as to maintain a harmonious balance between innovation and socio-economic considerations.

National Patent Strategy to promote Inclusive Innovation

Since the Uruguay Round Negotiations and the subsequent incorporation of the TRIPS agreement, the South is witnessing massive changes in their IP laws, leaving little space and elasticity for lawmakers. Still there are few flexibilities, which should be used by the South to address sensitive areas like public health, local industrial development, agriculture, environmental protection and IP laws should be drafted in a way that stimulate innovation in every sector. The role of IP varies depending on productive structures, sectors involved, levels of development and States' policies.

As the World Bank has noted, in the area of IP "one size does not fit all" (World Bank, 2001). WIPO has defined an 'IP Strategy' as a 'set of measures formulated and implemented by a government to encourage and facilitate effective creation, development and management of intellectual property. It outlines the manner in which to develop infrastructure and capacities to support inventors of IP to protect, develop and exploit their inventions' (Correa, 2008). It should also address the socioeconomic concerns of developing nations as subsequently emphasised by WIPO's 'development agenda'.

India's strategy should assure that the introduction or strengthening of patent protection for pharmaceutical products will not deny access to medicines to the bulk of its population and will enable its generic pharmaceutical industry to enjoy the TRIPS flexibilities (Musunga and Oh, 2006). Close scrutiny of patent applications is required, especially regarding the 'inventive' step in patent applications. This will it make it possible to reject frivolous patent claims. [7]

More than 60 percent of India's population depends on agriculture, which is supposed to play a critical role in India's inclusive growth and poverty alleviation. The TRIPS allows an effective *sui generis* system. It also allowed WTO members to exclude plants and animals from patentability (Tansey and Rajotte, 2008). [8] Many developing nations tend to adopt the UPOV Convention for the Protection of New Varieties of Plants following international pressure like FTAs. India's Plant Variety Protection and Farmers' Rights Act, 2001 acknowledge the role of rural communities and farmers' varieties in the breeding of new plant varieties (WTO: WT/GC/W/302). [9] India has accepted the diffusion of genetically modified varieties. Hence, patenting of genes and cells may have serious consequences for its agriculture (ISAAA, 2006). [10] To supplement a viable IP policy in agriculture, India should also select a better model and level of protection to its geographical indications.

Another area, which is gathering greater momentum in multilateral negotiations, is the issue of environment. The North is attempting to treat India and China at par with the developed nations. Both these nations are yet to develop an environmental-friendly clean technology. The TRIPS provision on transfer of technology is declaratory in character and imposes only moral obligation on its Member Nations. India did sense the danger of the creation of an oligopoly in patented environmental technology (WTO, 1997) [11] in 1997 itself. Provisions should be made to ensure access to environmental technology.

At the Committee on Trade and Environment, India stated: '...problem exists on its industry's access to certain patented technologies of Ozone Depleting Substances (ODS) substitutes required for implementation of the Montreal Protocol. [12] Naturally, a national patent strategy should attempt to ensure transfer of such technologies from the North. For instance, the developed nations went forward with the HFC technology and spent huge amounts on its research and development. Most of the multinational companies based in USA and in Japan adopted this technology in manufacturing compressor and refrigerator units. HFC 134a is under the patent protection of

Du Pont. Dow, ICI and Hoechst have acquired HFC 134a technology under cross licensing agreements. These technologies, which have been used since 1990-91 had gained a world-wide market share of approximately 40 percent by early 1996. When an Indian company sought the technology the price demanded was ten times more than the reasonable price (Wattal, 1998).

The strategic positioning of India as a technology superpower will also depend on its ability to make crucial technology acquisitions in the global market. Countries like Japan and Korea have successfully used such acquisitions to expand their IP portfolio and some Indian companies, especially in the pharmaceuticals and biotech sectors, have already been engaged in such acquisitions. However, such examples are sporadic and there is need for a national strategy on technology acquisition, with the aim of leapfrogging expertise in key areas. This is very much necessary for small scale industries that often fail to make technological updating and face extinction. Therefore, the National Knowledge Commission has suggested that the Central Government create a Global Technology Acquisition Fund (Knowledge Commission, 2007).

A high power expert committee body should be created in the National Knowledge Commission. The Committee should closely peruse the IPR dimensions of new and rapidly changing technologies, especially in ICT, biotechnology, nanotechnology, electronics, engineering and bio-informatics, among others. There is therefore, need for high-powered expert bodies that can help identify IPR issues arising out of each such areas with a view to evolving necessary IPR policies that would optimally foster greater global competitiveness for Indian industry as well as ensure faster innovation, wealth creation and overall development.

Being a nation with dual economy, a cross-cutting, multi-pronged strategy is needed to make India's innovation system work for the poor. But much of the discussions focus on R&D

of advanced sectors, leaving the masses out of the innovation system. Hence, an inclusive innovation system should be built up enabling the poor to reap the dividends of technological advancements without paying heavily for patents.

'Pro-Poor Intellectual Property Rights' that reward functional knowledge from traditional communities should be crafted. It needs bold experimentation in both legislation and administration. Bruce Abramson suggests three possibilities: (i) develop a direct national traditional knowledge analog to the patent system; (ii) institute a blanket license for traditional knowledge IPR; and (iii) establish a voluntary fund and encourage MNCs to contribute funds related to the value that they extract from the traditional knowledge (Abramson, 2007). [13] The IP policy that is sensitive to the agricultural interests of the country should also assess the mode and level of protection needed for geographical indications (GIs) that best suit the local conditions.

Patent laws are forcefully imposed on the South by the North. Hence, a sound national patent strategy for India also depends on its effective participation in international negotiations and its power to influence the IP norm-setting process in such deliberations.

End Notes

1. Interestingly, the only firms that named R&D as the most important area of focus were pharmaceutical firms. This emphasis on process is a sharp departure from traditional R&D and manufacture-centric notions of Innovation.
2. Indian drug maker Ranbaxy was taken over by Japan's Daiichi Sankyo in June 2008.
3. It is interesting to note that Samsung was a sub-contractor of Hitachi, but eventually it became a company capable of producing its own products.
4. Indian Law also provides a long list of exclusions from patentability (Section 3, Patents Act, 1970).
5. There is shortage of patent attorneys and patent examiners in the patent offices and also judges who are specially trained in patent law.

6. The debates that took place in the Indian legislature show that the hearts of the parliamentarians, irrespective of their party lines, were all against amending the Patent Act, 1970, which was framed in such a way wherein "medical discoveries would be free of patents and there will be no profiteering from life or death". But still the Parliament was forced to pass legislation in a hurry to abide by the WTO's determination to amend the Indian patent law before April 1999, which manifested that WTO has become a *de facto* world government in respect of commerce and trade.
7. Recently, in the Novartis case the Supreme Court suggested that the Government should consider having an expert member on the Intellectual Property Appellate Board (IPAB) that reviews decisions of the Indian patents office.
8. Article 27.3(b) of the TRIPS Agreement; See *Geoff Tansey and Tasmin Rajotte* (eds.) (2008), *The future control of food. A Guide to International Negotiations and Rules on Intellectual Property, Biodiversity and Food Security,* Earthscan, Ottawa-Sterling (USA).
9. Kenyan submission on behalf of the African Group at the WTO General Council is a nice addendum in this regard. It suggested that the following provisions should be included for any *sui generis* system: (i) the protection of the innovations of indigenous and local farming communities in developing countries, consistent with the Convention on Biological Diversity and the International Undertaking on Plant Genetic Resources; (ii) the continuation of the traditional farming practices, including the right to save, exchange and save seeds, and sell their harvest; (iii) preventing anti-competitive rights or practices which will threaten food sovereignty of people in developing countries, as is permitted by Article 31 of the TRIPS Agreement . See WTO Doc. WT/GC/W/302, General Council - Preparations for the 1999 Ministerial Conference-The TRIPS Agreement-Communication from Kenya on Behalf of the African Group.
10. Argentina, Brazil, India and China have adopted the genetically modified crops globally; Mexico, Philippines, Paraguay, Uruguay, Honduras, Colombia, South Africa and Iran have begun cultivation of such crops. See *ISAAA Brief 35-2006: Executive Summary, Global Status of Commercialized*

Biotech/GM Crops: 2006, available at, www.isaaa.org/resources/publications/briefs/35/executivesummary/default.html.

11. See WT/CTE/W/66, November 1997.
12. This was confirmed by the intervention made by the Republic of Korea in the CTE. In an UNCTAD Meeting on positive measures held in November 1997, the People's Republic of China also confirmed that it was facing similar problems on accessing these patented technologies. At this meeting, representatives of UNEP and the Multilateral Fund confirmed that such problems had been brought to their notice and that so far no guidance has been finalized for funding projects on the production of ODS substitutes in these countries. See WTO Doc. WT/CTE/W/82, April 1998.
13. Bruce Abramson (2007): "India's Journey toward an Effective Patent System." *Policy Research Working Paper 4301,* South Asia Finance and Private Sector Development, World Bank, Washington, DC. CSIR, India's premier research agency, make significant efforts in this regard by fostering rural technology. See CSIR (Council of Scientific and Industrial Research), "Making Technology Work for Rural India: A Step towards Rural Transformation" New Delhi, 2004 available at, http://www.csir.res.in/csir/External/Utilities/Frames/achievements/main_page.asp?a=topframe.htm&b=leftcon.htm&c=/csir/external/Hed/achievements/rural_technologies.htm. Also see Mashelkar, R. A. 2001. "The Indian Innovation System", Development Outreach, World Bank, Washington, DC.

References

Abramson, Bruce (2007): "India's Journey toward an Effective Patent System", *Policy Research Working Paper 4301,* South Asia Finance and Private Sector Development, World Bank, Washington, DC. CSIR, India's premier research agency makes significant efforts in this regard by fostering rural technology.

Basheer, Shamnad (2007): "India's New Patent Regime: Aiding Access or Abetting Genericide?", 9 *International Journal of Biotechnology, 122.*

Bird, Robert C. and Daniel R. Cahoy (2007): "The Emerging BRIC Economies: Lessons from Intellectual Property Negotiation and Enforcement", 5 *North-western Journal of Technology and*

Intellectual Property, 400.

Bowonder, Kelkar, N.G. Satish, J.K. Racheria (2006): "Innovation in India: Recent Trends", *TTMC Research Paper,* March 31.

Colin, Jeffrey (2007): "Coming into Compliance with TRIPS: A Discussion of India's New Patent Laws", 25 Cardozo Arts and Ent LJ 877.

Correa, Carlos M. (ed.), (2008): *A Guide to Pharmaceutical Patents Vol. I & II,* Geneva: South Centre, July.

CSIR (Council of Scientific and Industrial Research), (2004): "Making Technology Work for Rural India: A Step towards Rural Transformation", New Delhi, available at, http://www.csir.res.in.

Fan, Peilei (2008): "Innovation Capacity and Economic Development: China and India", *Research Paper No. 2008/31,* United Nations University, World Institute for Development Economics Research.

Gupta, Rishi (2005): "TRIPS Compliance: Dealing with the Consequences of Drug Patents in India", *Houston Journal of International Law,* 599-602.

Herstatt, Cornelius, Rajnish Tiwari, and Stephan Buse (2008): "India's National Innovation System: Key Elements and Corporate Perspectives", *TIM/TUHH Working Paper 51,* January.

ISAAA Brief 35 (2006): *Executive Summary, Global Status of Commercialized Biotech/GM Crops*, available at, www.isaaa.org/resources/publications/briefs/35/executivesummary/default.html.

Mani, Sunil (2004): "Institutional Support for Investment in Domestic Technologies: An Analysis of the Role of Government in India", *Technological Forecasting and Social Change*, 71 (8): 855-863.

Mani, Sunil (2007): "Government Support for Sustaining a Knowledge-based Economy: An Examination of India's Technology Policy Initiatives, 1990-2005", in A. Vaidyanathan and K. L. Krishna (ed.), *Institutions and Markets in India's Development-Essays for K.N. Raj*, New Delhi: Oxford University Press.

Mashelkar, R.A. (2001): *The Indian Innovation System,* Washington,

DC: Development Outreach, World Bank.

Mathew, James and Gireesh C. Prasad (2005): "Ranbaxy-Cipla in talks with Roche for Tamiflu Deal", 13th December, *Economic Times,* http://economictimes.indiatimes.com/articleshow/1329036.cms>

Musungu, Sisule and Cecilia Oh (2006): *The Use of Flexibilities in TRIPS by Developing Countries: Can they Promote Access to Medicines?*, Geneva: WHO, South Centre.

National Knowledge Commission *Recommendations on IP* (5 January 2008): available at, www.knowledgecommission.gov.in/downloads/reccomondations/enterprenershipPM.pdf National Knowledge Commission, Innovation in India, June 2007.

Orazco, David (2007): "Will India and China Profit from Technological Innovation?", 5 *North-western Journal of Technology and Intellectual Property*, 426.

Osenga, Kristen (2007): "Closing in On Open Science: Trends in Intellectual Property and Scientific Research: Rembrandts In The Research Lab: Why Universities Should Take A Lesson From Big Business To Increase Innovation", *59 Maine Law Review,* 40-42.

Rai, Rajnish Kumar (2008): "Scare of Avian Flu Revisits India: A Bumpy Road Ahead", 8 PGH, *Journal of Technology Law and Policy,* 7.

Utz, Anuja and Carl Dahlman (2007): "Promoting Inclusive Innovation", in *Unleashing India's Innovation: Toward Sustainable and Inclusive Growth,* Mark A. Dutz (ed.), World Bank, available at, www.sitersources.worldbank.org/southasiaext/resources/223546 1181699473021/3876782/indiainnovtionchapter4.pdf.

Watal, Jayashree (1998): "The Issue of Technology Transfer in the Context of the Montreal Protocol: Case Study of India", in Veena Jha and Ulrich Hoffmann (eds.), *Achieving Objectives of Multilateral Environmental Agreements: A Package of Trade Measures and Positive Measures. Elucidated by Results of Developing Country Case Studies*, UNCTAD/ITCD/TED/6, Geneva, UNCTAD.

World Bank (2001): *Global Economic Prospects and the Developing*

Countries 2002, Washington, D.C.

World Bank (2007): *Unleashing India's Innovation: Towards Sustainable and Inclusive Growth,* World Bank.

WTO Docs (1998): WT/GC/W/302, General Council-Preparations for the 1999 Ministerial Conference-The TRIPS Agreement-Communication from Kenya on Behalf of the African Group, footnote; WT/CTE/W/66, November 1997; WTO Doc. WT/CTE/W/82.

6

The Trouble with SAARC

Harsh V. Pant

Established in 1985, the South Asian Association for Regional Cooperation (SAARC) had its 16th summit meeting in Thimpu, Bhutan in 2010. Apart from the fact that Bhutan hosted its first SAARC summit, there was hardly anything that inspired confidence in this largely moribund organization that also celebrated the 25th anniversary of its founding this year. Covering at least 1.5 billion people across India, Pakistan, Bangladesh, Nepal, Sri Lanka, Bhutan, Maldives and Afghanistan, SAARC is one of the largest regional organizations in the world. But its achievements so far have been so minimal that even its constituents have become lackadaisical in their attitudes towards it. The state of regional cooperation in South Asia could be gleaned from the fact that to attend the summit, Pakistani Prime Minister Yousaf Raza Gilani reached Bhutan via Nepal, using Chinese territory in Tibet rather than the straightforward route through India.

Background

In November 1980, President Ziaur Rahman of Bangladesh sent a working paper on Regional Co-operation in South Asia to his fellow leaders in the region setting out the objectives of the South Asia Forum. The proposal revealed an understanding of the bilateral issues among the states and very cleverly and deliberately omitted political and security matters. Areas identified for regional cooperation were non-political and non-security in nature, and included telecommunications, tourism, agriculture, transport and meteorology (Paranjape, 2002: 348). No rigid time frame was proposed, however it was envisaged that a dialogue process would be initiated.

The proposal was finally developed into a declaration with the formation of the SAARC at the Dhaka Summit in Bangladesh in 1985.

With Bangladesh, Bhutan, India, Maldives, Nepal, Pakistan and Sri Lanka as its original members, the SAARC was established in 1985 to promote the welfare of the people of South Asia; to accelerate economic growth, social progress and cultural development in the region; to promote collective self-reliance; to promote collaboration in economic, social, cultural and technical area; and to strengthen cooperation among member states in international forums on matters of common interest. Afghanistan became SAARC's eighth member in 2007. The US, EU, China, Japan, South Korea, Iran, Myanmar, Mauritius, and Australia have been granted observer status in the last few years.

To achieve its objectives several steps have been taken to give institutional shape to SAARC. A permanent secretariat, headed by a Secretary General appointed by member states in alphabetical order for a three-year term, was established in Kathmandu in 1987. The highest authority of the association rests with the Heads of State or Government. The next level of decision making is the Council of Ministers which comprises of the Foreign Ministers of the member states who formulate the policies, review progress, decide on new areas of cooperation and decide on general matters of interest to the organization. SAARC's ultimate objective, in fact, is not merely to increase regional cooperation in the economic sphere, but by adopting this functional approach also to contribute to mutual trust and understanding among the member states, which would lead to cooperation among them in the political sphere too.

Yet while other regional associations—most notably the Association of South East Asian Nations (ASEAN) and the European Union (EU)—have grown economically by leaps and bounds, SAARC has made little progress on the high ideals promised in its charter. At the time of the proposal, the world was witnessing the Cold War and the region was divided between the power houses. Initially India and Pakistan were not even responsive to the formation of such an association. India saw it as a ploy to reduce Soviet influence in the region by ganging up against the Soviets since they were in Afghanistan whereas Pakistan feared that this would legitimize India's economic dominance in the region (Kumar, 2005). SAARC follows the principle that all decisions have to be made unanimously and that no bilateral and contentious issue can be on the organization's agenda. The result is even though SAARC has been in existence for over two decades now, it has not been able to present itself as a formidable organization on the global platform.

Economic Impediments in South Asia

Half of the world's poor continue to live in South Asia, although it constitutes only three percent of the world's area. Trade among SAARC states has remained limited despite the fact that all are located within a close proximity of one another and are all part of the World Trade Organization (WTO). Intra-regional trade among SAARC states is way below potential and does not involve goods vital to the economics of South Asia states. Smaller States in the region are haunted by the fear of Indian goods sweeping their markets.

Only 3.2 percent of Bangladesh's trade is with the region and only Sri Lanka has a Free Trade Agreement with India. Pakistan does not even give the WTO-mandated Most Favoured Nation (MFN) status to India. Smuggling continues to overshadow legal trade between India and Pakistan. Despite the telecommunication connectivity, intra-trade and investments flow are far less compared with East and South East Asia.

There are a number of economic impediments to regional economic cooperation in South Asia. The size of the markets is highly unequal. India has the biggest market with 76 percent of the population, 59 percent of the import market, 62 percent of the export earnings, 71 percent manufacturing and 69 percent manufacturing exports (Chaudhury, 2006). The economies of the other member states are much smaller. The fear of domination by India definitely hampers trade despite categorization of member states into Least Developed Countries (LDCs) and non-LDCs.

There is a lack of product diversity. The market in South Asia is very competitive rather than complementary because of the nature of the exportable commodities. Textiles account for major exports from Bangladesh and Pakistan, tea form major exports from Sri Lanka and India, jute is shared between India and Bangladesh, cotton and leather are exported from India and Pakistan. Thus, apart from limited items for export competition exists since similar commodities form part of the exportable products.

High tariff rates and tariff barriers encourage informal trade between the countries. In simple terms this is smuggling, which happens unofficially and the degree has kept increasing over the years. It is estimated that unofficial trade between India and Bangladesh only is in excess of US$ 1.5 billion (Chaudhury, 2006: 184-85). Some of the trade between India and Pakistan happens through third countries like Dubai, since tariffs are high. Machinery used in textile and leather industry in Pakistan is imported into Pakistan from India via Dubai.

Poor transportation and communication system in the region result in reduced trade within the region. The SAARC nations lack a well developed road and rail network. Despite the continuous landmass all through except Sri Lanka, no effort has seriously been undertaken to have an Asian Highway or rail network for free and smooth flow of goods across the region. No regional shipping lanes exist to facilitate transit of goods and people across the region. Stringent policies for visas and immigration for people as well, denies opportunity of tourism and for business travel.

Thus there are major challenges which restrict deeper

economic integration between the South Asian countries. Trade among the countries is restrictive due to lack of infrastructure and institutionalization of sectoral reforms. The comparative benefits lies in low technology products such as leather, clothing, textiles, footwear and agricultural products where there is still competition amongst members. Apprehensions of the smaller economies of losing out to larger ones are resulting in the continuation of high tariff rates to safeguard domestic industry and farmers. Increasing cross border illegal trade needs to be checked to enhance chances of additional trade benefits within the region.

To facilitate healthy trade within the region the critical ingredients are a good rail and road network, as well as an advanced banking services, all of which need to be instituted. A well developed road and rail network connecting the entire region along with good communications would go a long way in increasing trade and the much needed tourism. A developed banking sector would also be instrumental in facilitating trade negotiations among the corporate sector. Friendly taxation laws for foreign firms and not being overprotective of the domestic industry would invite foreign investment into the region which has abundance of cheap and skilled labour as is happening with the services sector in India.

It is not surprising then that SAARC shunned economic integration until 1995 when the member states decided to launch the South Asian Preferential Trading Arrangement (SAPTA) to promote economic cooperation within the region. For almost a decade, SAPTA meandered along an unproductive course of commodity-by-commodity negotiations for reducing tariffs until the South Asian free Trade Agreement (SAFTA) was launched in 2004 (Chaudhury, 2006).

Thus, South Asia which as it is was a late starter took a long time in institutionalizing regional trade, which blocs like EU and ASEAN successfully achieved much earlier in their institutional trajectory. SAFTA went into force in 2006 creating a framework for the establishment of a free trade area covering 1.4 billon people. The agreement calls for zero customs duty on the trade of most goods in the region by the end of 2016. India and Pakistan are yet to ratify the treaty (Chowdhary, 2007).

SAFTA plans to boost the present five percent trade by phasing out tariff barriers and quota restrictions for goods between member countries over a specified timeframe. The LDCs namely Nepal, Bhutan and Maldives will get 10 years or until 2016, to reduce their tariff barriers to zero to five percent, whereas the non-LDCs namely, India, Pakistan and Bangladesh will get five years to implement the tariff regime and Sri Lanka six years (Dubey, 2007).

But political clashes between South Asian countries have ended up jeopardizing the effectiveness of regional trade agreements. They also lead individual SAARC nations to advance their economic interests through bilateral agreements, reducing the incentive to engage multilaterally. Much of the trade that does occur in South Asia has been facilitated by bilateral arrangements among countries (Dubey, 2007). If the trend continues, then this will further erode SAARC's credibility with states having less incentive to pursue economic cooperation through SAARC.

Political Impediments in South Asia

Despite an extraordinary degree of cultural affinity in South Asia, it is also a region mired in mutual distrust. Bilateral disputes among member states have undermined regional efforts to promote trade and economic cooperation. India's predominant power position in South Asia has allowed conflicts between India and its neighbours to undermine organizational unity. Due to political tensions among the member states, SAARC hasn't had a very glorious record and has had only 17 Summit level meetings in the last 26 years where it was meant to have one each year. Territorial disputes and other political issues have had a few meetings either postponed or cancelled.

Problems first started to surface when the fifth summit in 1989 had to be postponed due to tensions between Sri Lanka and India over the role of Indian Peace keeping Force (IPKF). These issues were certainly bilateral and political in nature but did they not affect the region? Similar issues have emerged subsequently between India and Pakistan when India refused to attend the summit after the attack on its parliament and the stand off between the countries in 2002. India-Pakistan squabbles tend to hijack regional initiatives, to the irritation of smaller neighbours.

India-Pakistan conflict is at the heart of SAARC's dysfunctionality. Pakistan has traditionally insisted that the dispute over Kashmir needs to be resolved before it can agree to move forward economic cooperation with India. Islamabad has refused to grant India the MFN status as required by its WTO membership. The two countries have fought four bloody wars since partition, the last one being the Kargil War in May 1999. This was followed by a military confrontation for over seven months in 2001-2002 in the aftermath of an attack on the Indian Parliament in December 2001. Terrorists trained in Pakistan launched attacks on Mumbai in November 2008, leading to a freeze in Indo-Pak ties (Ganguly, 2002).

One way India has tried to overcome this problem is by trying to improve its ties with the rest of South Asian states. The former Prime Minister of the country, I.K. Gujral, even enunciated a doctrine—the Gujral doctrine—which talked about India trying its best to overcome the suspicions of other regional states by being "generous" and giving unilateral concessions to its smaller neighbours (Gujral, 2003). And as a result India signed 30-year water sharing treaty with Bangladesh and a trade and transit treaty with Nepal.

A sub regional grouping also evolved under the SAARC including Bangladesh, Bhutan, Nepal and India. Other more effective alternatives to SAARC have also started emerging. The Bay of Bengal Initiative for Multi-Sectoral Technical and

Economic Cooperation (BIMSTEC), originally involving Bangladesh, India, Sri Lanka, and Thailand, was established in 1992. Bhutan and Myanmar joined in subsequent years making this forum almost a buffer zone between the more effective ASEAN and SAARC itself.

Rising Profile of China in South Asia

For long, the dominant narrative of SAARC has been how the India-Pakistan rivalry hampers its evolution into anything of significance. That is now rapidly losing its salience with China's growing dominance of the South Asian landscape. China entered SAARC as an observer in 2005, supported by most member states. Despite its misgivings, India could do little about it and so had to acquiesce. Now, much to India's consternation, Pakistan, Bangladesh and Nepal are supporting China's full membership in SAARC.

China's rising profile in South Asia is no news. What is astonishing is the diminishing role of India and the rapidity with which New Delhi is ceding strategic space to Beijing on the subcontinent. Even as China is becoming the largest trade partner of most states in South Asia, including India, New Delhi's strategic hold on South Asia is weakening.

Pakistan's 'all-weather' friendship with China is well-known, but the reach of China in other South Asian states has been extraordinary (Pant, 2011). Bangladesh and Sri Lanka view India as more interested in creating barriers against their exports than in spurring regional economic integration. India's protectionist tendencies have allowed China to don the mantle of regional economic leader. Instead of India emerging as facilitator of socioeconomic development in Sri Lanka, Nepal and Bhutan, it is China's developmental assistance that's having a larger impact.

India's attempts to keep China out of the subcontinent have clearly not worked, and it's time to re-evaluate its South Asia policy. China's strategy towards South Asia is premised on encircling India and confining her within the geographical coordinates of the region. This strategy of using proxies started with Pakistan and has gradually evolved to include other states in the region, including Bangladesh, Sri Lanka and Nepal. It is entering markets in South Asia more aggressively through both trade and investment, improving linkages with South Asian states through treaties and bilateral cooperation. Following this up by building a ring of road and port connections in India's neighbourhood and deepening military engagements with states on India's periphery, China has firmly entrenched itself in India's backyard.

This quiet assertion of China has allowed various smaller countries of South Asia to play China off against India (Pant, 2010). Most states in the region now use the China card to balance against the predominance of India. Forced to exist between their two giant neighbours, the smaller states have responded by a careful balancing act.

India's structural dominance in South Asia makes it a natural target of resentment among its smaller neighbours. India's looming presence in the region makes for the unique paradox where the core regional power itself is seen as constituting a security threat. There are major disputes within the region and somehow India seems to occupy centre stage. Most states have bilateral disputes with India. Bangladesh remains concerned about India exploiting its geographical position to redirect water flows while India's control over their transit links makes Nepal and Bhutan susceptible to pressure from New Delhi.

These disputes have led to a climate of political animosity and suspicion which has resulted in limited cooperation. A power imbalance among states makes it difficult for a regional organization to work effectively. India in South Asia is far ahead of all the regional states in its economic power, military might as well as global influence. This makes other small states in the region apprehensive of India's intentions. While they recognize the importance of India in facilitating faster economic growth in the region, they are also reluctant to work with India for fear that India will use SAARC to further its "hegemonic ambitions."

Yet there is no hope for regional economic cooperation in the absence of Indian leadership. India is likely to be the driving force of South Asian economy in the years to come, and rather than being apprehensive, the neighbouring countries must take advantage of this situation. If South Asian countries can cooperate and take advantage of India's developed market and technological supremacy in the field of information technology, biotechnology, pharmaceuticals and in the auto sector, they will only be helping themselves.

The failure of India in countering China's rise has made it even more unlikely that such cooperation will evolve productively. As the two regional giants compete with each other in the near future, they will be more focused on their relative gains vis-à-vis each other than in the absolute gain that regional cooperation can bestow.

Limiting Expectations

There is little likelihood of SAARC being able to focus on security issues like the ASEAN Regional Forum (ARF) in the near future given the plethora of bilateral disputes among member states and the sensitivities involved. In its 25 year long history, SAARC has been a mute spectator to the myriad security issues in the region—India-Pakistan tensions, Maoist threat in Nepal, militarization of politics in Pakistan and Bangladesh, and Sri Lanka's struggle with the Tamil separatists. Paradoxically, as India becomes more powerful and ambitious in projecting its power abroad, the more reluctant its smaller neighbours will be in accepting its leadership in the region. And so their dependence on China is likely to increase even further.

In the short to medium term, the best option for the region is to make SAARC a more effective economic organization. Towards this end, focus should be on improving connectivity as a means of enhancing regional integration. A multi-nodal transportation system, air connectivity, relaxing visa regulations, intra-regional tourism, and academic exchanges are some means of improving connectivity. Efforts should be made to reach out to the civil society and the private sectors to supplement the role of the respective governments. SAFTA needs to be implemented effectively and swiftly with the participation of all member states at all levels. SAARC's outreach to external actors, in particular other regional organizations should be encouraged so that learning can take place.

Liberals in South Asia have long taken their inspiration from the extraordinary developments in the European Union (EU), arguing that South Asia can also go down a similar path of regional economic and political cooperation. However, that's a fundamentally flawed comparison for a number of reasons not the least because the states in Western Europe could arrive at the EU only after resolving their persistent security dilemmas. And the US security umbrella continues to ensure that European political rivalries do not raise their ugly heads again. In South Asia, the security dynamics between a large India and its smaller neighbours ensures that the road to economic and political cooperation will be a bumpy one. And after the emergence of China, that road will be an even more difficult one traverse.

The region is facing an unexpected challenge today from terrorism, extremism, poverty, climate change, narcotics and organized crime. Progress in SAARC has stalled due to hesitancy borne from historical legacies, differences and disputes. The member states are unwilling to invest in SAARC and public perception of the organization is largely negative.

The Indian Prime Minister Manmohan Singh rightly put it across in his address during the 14th SAARC Summit held in New Delhi in 2007:

"To realize our hopes, we need SAARC to be an efficient instrument implementing what we member states seek. After several years of effort, the time has come to move SAARC from a declaratory phase to action and implementation."

This is important if SAARC is not to become a symbolic forum with little regional relevance. However, at the moment there is little likelihood of this happening any time soon.

References

Chaudhury, Anasua Basu Ray (2006): *SAARC at Crossroads: The Fate of Regional Cooperation in South Asia*, New Delhi: Samskriti, pp. 181-182.

Chowdhary, Mahfuz R (2007): "The Economic Integration of SAARC", July 29, available at, http://www.americanchronicle.com/articles/view.

Dubey, Muchkund (2007): "SAARC and South Asian Economic Integration," *Economic and Political Weekly*, April 7, pp. 1235-1238.

Ganguly, Sumit (2002): *Conflict Unending: India-Pakistan Tensions Since 1947*, New Delhi: Oxford University Press.

Gujral, I.K. (2003): *Continuity and Change: India's Foreign Policy*, New Delhi: Macmillan.

Kumar, Ranjit (2005): *South Asian Union: Problems, Possibilities and Prospects,* New Delhi: Manas Publication, pp. 73-75.

Pant, Harsh V. (2010): *The China Syndrome: Grappling with an Uneasy Relationship*, New Delhi: HarperCollins.

Pant, Harsh V. (2011): *China's Rising Global Profile: The Great Power Tradition* (Portland, OR: Sussex Academic Press, pp. 29-50.

Paranjape, Shrikant (2002): "Development of Order in South Asia: Towards a South Asian Association for Regional Cooperation Parliament," *Contemporary South Asia*, p. 348.

7

Indo-US Trade Relations

Shaijumon C.S.

India is a large emerging economy, a key driver of the global economy, and a major player in Asia. India, the world's 4th largest economy after US, China and Japan, has per capita income on a PPP basis was US$ 2,780, equal to only 7.4 percent of US levels. According to World Bank, India has 433 billion people (44.2 percent of the population) living below the international poverty measurement of less than US$ 1 per day. But Indian estimates say that the poverty rate is 26 percent of its population. Growing debt situation of US with China, the US subprime crisis in 2007 and more recently the debt problem of United States of America have challenged the supremacy of US economy in the global arena. Considering that India, a country with a GDP of around US$ 1 trillion, accounts for a mere 1.3 percent of US trade, there is substantial potential to increase bilateral trade. This paper argues that the world's oldest and largest democracies should engage in more economic activities for an inevitable global balance of trade.

Most of India's independent history, the US remained its foremost trading partner, a primary source of investments, the first resort for technology acquisition, and significant in funds and aid support. India was much influenced by American ideas of human freedom and democracy in drafting its constitution. But relations estranged during the time of Cold War because of divergent foreign policy views. The visit of Prime Minister Rajiv Gandhi to US in 1988 was a landmark in the economic relationship of both the countries, which laid the foundation for India's knowledge economy, and reinstated the US as a preferred economic partner.

India's sweeping economic reforms in 1991, together with the collapse of the Soviet Union, and a new desire for strengthening relations between the two world's strongest advocates of democracy, set the stage for a new paradigm in economic transactions. Again the relations got estranged in the conduct of India's nuclear test in 1998, which has led to a set-back in economic relations, but the President Clinton's visit in 2000 gave a new life the economic relationship of both the countries. President Bush was eager to continue the constant economic relationship with India and which culminated in the Nuclear Energy Agreement. Now the two national economies have complimentary business interests, rather than a standard developed-developing country engagement in their mutual economic relations.

Table 7.1: Selected Indicators of USA, India and China

	USA	India	China
Total population	31 crore	121 crore	133 crore
Population density	31/km² (80/sq mi)	356/km² (922/sq mi)	110/km²
GDP (nominal)	US$ 14.256trillion (US$ 46,381 per capita)	US$ 1.430 trillion (US$ 1,176 per capita)	US$ 4.99 trillion (US$ 4,382 per capita)
Indian Americans	27.66 lakh Indian origin people in US	60,000 American born people living in India	
Military expenditure	US$ 663.7 billion	US$ 37.6 billion	US$ 91.5 billion

Source: Wikipedia.

The most striking difference in the Indian and Chinese demographic pictures over the coming decades is the onset of India's youth bulge at the same time that China finds its population graying. U.S. Census Bureau analysts estimate that new entrants into China's labour force may be near its upper limits of 124 million as the population of Chinese aged 20 to 24 peaks this year. India's population of 20- to 24-year-olds, on the other hand, is not expected to peak until 2024 when it hits 116 million. While India's workforce will increase by 110 million over the next decade, China's will increase by less than 20 million, according to a Goldman Sachs study.

This demographic dividend could fuel India's economy in ways that make it a peer competitor to China—in particular, pushing Indian growth rates ahead of China's. At present, the Chinese economy is vastly larger than India's. At more than US$ 4.7 trillion, China's GDP is four times India's; its GDP per capita, at about US$ 3,565, is three times India's; and China produces about 12 percent of the world's GDP while India produces about 5 percent. The Chinese also hold socioeconomic advantages over India that could play in Beijing's favour: Adult literacy in China stands at about 91 percent, compared to roughly 61 percent in India.

Trade relations

India was the world's 30th largest merchandise exporter and the 24th largest importer. India's principal exports in 2001 were textiles (22 percent), gems and jewellery (16.8 percent) and chemical and related products (14.5 percent). Its top imports were petroleum (27.4 percent), pearls and precious and semi precious stones (9 percent) and gold and silver (8.9 percent). India's top three export markets were the US (19.5 percent), the UAE, and Hong Kong, and its top source for imports were the US (6.1 percent), Switzerland and Belgium.

The total volume of bilateral trade between India and US was only US$ 4 billion in 1986. By 2006, the total value had raised to US$ 31.9 billion—nearly an eight fold increase. When we look up the history of Indo-US trade relationships, both India's exports to and imports from the United States were relatively low in value and subject to fluctuations. In the mid 1960's nearly one-fifth of India's exports went to the US, and the US supplied India with over one third of its imports. In 1981, the US had declined in importance, purchasing just over 11 percent of India's exports and supplying less than 10 percent of its imports. In 2006, the US purchased 17.4 percent of India's exports (almost same percentage as in 1961) but only provided India with 5.9 percent of its imports (Table 7.2).

Table 7.2: Volume of Trade between India and US

Year	Percent of Total Exports to USA	Percent of Total Imports from USA
1961	20	33
1981	11	10
2006	17	5.9
2010	11	5.9

Interestingly both the percentage share of imports from and the exports to US has been on a declining trend. This means the importance of the American economy in the trade relationship of India is declining. The situation is really a thought provoking one, because of the importance that US and India are having in the world economy. United States as the largest economy in the world and India as the home of second largest market in the world, more engagement between these countries has to happen for a global balance of trade relations. When we look in depth on the composition of products that is being trading between countries, we will understand the kind of potential left behind for further engagement of both the countries and thereby creating more value addition. The top US exports to India in 2006 included machinery, electrical machinery, optical and medical instruments, and aircraft. The top Indian exports to US in 2006 included jewellery, woven apparel, knitted apparel and miscellaneous textile articles.

Both nations need to recognize that China is a rising competitor to the Indo-US merchandise trade. China is now one of the top three trading partners of India whereas India is one of top ten trading partners of China. China's trade is increasing much faster with India out of top ten trading partners and same is in the case with India (Pillania, 2010). Table 7.3 clearly depicts the importance of China and USA in the economy of India.

Table 7.3: China and US in the Economy of India

	China		USA	
	2000	2010	2000	2010
Indian exports (US$ billion)	1.05	11.6	9.3	19.5
percent of Indian exports	0.5	6.5	20	11
Indian Imports (US$ billion)	1.5	30.8	2.9	17
percent of Indian imports	0.6	11.9	6	5.9

It is a fact that there has been a recent rapid increase in bilateral merchandise trade between India and USA. However, despite the rise in the value of bilateral trade, the relative importance of the other country to the nation's external trade volume has remained small and is well below the levels that seen at the time of independence. As we had seen, over the last five years, India's trade with China has grown rapidly than trade with the US. As a result, China has already surpassed the US as India's source of imports, and may soon become India's largest trading partner. China's and India's economic growth rates have outperformed world average growth rates for the most part of last fifteen years. China has grown at an average rate or close to ten percent, whereas India has grown at an average rate of six percent annually (Table 7.4).

India's economy (GDP on PPP basis) was about three quarters the size of China's, but by 2003 it fallen to 44 percent China's size. India's living standards were slightly greater than China's in 1990, but by 2003 it had fallen to 54 percent of China's. India's exports relative to Chinese exports fell from 29 percent in 1990 to 12 percent in 2003, while imports dropped from 55 percent to 17 percent. India make small gains in FDI flows relative to China over this period, however, the total level of FDI stock in China remains substantially higher than in India. This widening gap between India and China to differences in the pace and scope of economic its trade and investment regimes. For example, China's average tariff has fallen from 43 percent in 1992 to 12 percent in 2002. India's average tariff during this period dropped substantially, from 128 percent to 32 percent, but still remains among the highest in the world.

Table 7.4: Selected Comparative Indicators of India and China

	India		China		India's Size Relative to China (percent)	
	1990	2003	1990	2003	1990	2003
Population (millions)	850	1,062	1,139	1,295	74.6	82
GDP, PPP basis (US$ billion)	1,189	2,951	1,583	6,675	75.1	44.2
Per Capita GDP in US$ PPP	1,400	2,780	1,390	5,150	100.1	54.1
Exports (US$ million)	17,975	54,000	62,090	438,500	28.9	12.3
Imports (US$ million)	23,438	68,800	42,354	413,098	55.3	16.7
FDI Stock (US$ million)	1,592	34,559	68,513	490,243	2.3	7

Source: CRS Report for Congress, The Library of Congress, February 25, 2004.

In 2003, India was the 24th largest US export market and

the 18th largest supplier of US imports. India's exports to USA doubled between 2000 and 2007-08 to US$ 20.7 billion. Imports from USA rocketed up from US$ 3.7 billion in 2000 to US$ 21 billion in 2007-08. Total trade nearly tripled from US$ 14.3 billion in 2000 to US$ 41.7 billion in 2007-08. In 2006, USA continued to be India's premier destination for exports, with 17.4 percent share of India's exports.

However, its share in India's exports declined to below 12 percent by December 2008. US position in India's imports came down to fourth by December 2008. India accounted for 1.2 percent of US imports and 1 percent of its exports in 2006. India continues to be USA's 21st largest export market and 17th largest import supplier. 61 percent of India's software exports were to US in 2006-07. Services trade between India and USA stood at US$ 20 billion in 2007. US exports to India stood at US$ 9.4 billion, while Indian exports to USA were at US$ 9.6 billion. Total trade in 2007 was almost double the figure in 2005 (US$ 10.2 billion). About 27 percent of its computer and software requirements come from India, its largest source. India ranks as 10th among all countries as a source of services for USA. China is now India's largest trade partner, displacing USA. Thus, both countries trade with China has been growing rapidly, indicating that India and USA have unexplored potential in their trade (Table 7.5).

Table 7.5: India and United States Trade with China

	2000	2007-08
US exports to China (US$ billion)	16.3	65.2
US imports from China (US$ billion)	71.2	321.5
India's total trade with China (US$ billion)	3	10.8

Decadal Analysis of US Trade in Goods with India

For the simplicity of analysis, this paper is analysing the situation of Unites States trade in goods with India for the last twenty five years from the year 1985. As we are aware, the decade 1980's has played a significant role in spurring growth of the country in post liberalisation era. Aravind Panagariya (2004) has rightly pointed that, 'the difference between the reforms in the 1990's and those in the 1990's is that the former were limited in scope and without a clear road map whereas the latter were systematic and systematic".

The government was aggressively pursuing for policies to break the shackles of the 'contagious Hindu Rate of Growth' of Indian economy. Pre-reforms situation in the Indian economy has shown not much effect in the trade relationship between US and India. The volume of trade between countries has not increased much as expected. India was experiencing a favourable trade balance with United States throughout this period. Theoretically speaking, a favourable trade balance is good for the economy of a country. But it should not be at the cost of the domestic development by using the imported inputs.

In the next decade 1990's, India has had whole liberalising reforms, opening itself to foreign investment and trade and has ripened the benefits of unprecedented economic growth. The substantial reforms of the 1980's gave way to more systematic and deeper reforms of the 1990's. India was actually looking for an important space in the emerging global economy through the introduction economic reforms. It is important to have greater access to US economy for reaping the maximum benefits from the growing pie of economic growth in the world economy. The July 1991 reforms did away with import licensing on virtually all intermediate inputs and capital goods. Tariff rates have been reduced substantially.

However, the situation did not improve in the decade 1990's either. The volume of import and exports did not make a significant increase. But from July 1997 to July 2000, there were some improvement in volume of US imports from India. This might be because of our aggressive promotion of exports to various destinations. Booming bilateral trade has come to be the strongest pillar of China-India rapprochement in 1990's. This has not only since overtaken the pace of political confidence-building but also has a substantial impact on their mutual perceptions.

The benefits of the concerted efforts of economic reform of India during the decade of 1990 have started reaping actual benefits during the decade of 2000's. The volume of trade between US and India has increased. The hangover affected in Indo-US trade relations because of India's nuclear test in 1998 started fading away. The necessity of rapprochement between both countries in the wake of China's aggressive attitude in the world economy has been felt and which has paved the way for increase in bilateral trade between the countries.

From January 2001 to January 2001, the exports from US to India were more or less stable and hassle free. But from June 2007 to January 2010, there was a clear indication of fluctuation of exports of US to India. This was primarily because of the Sub-prime crisis experienced in US and thereby it affected the export financing of US industries. Indo-US trade, which shot up almost three-fold from US$ 13.7 billion in 2001 to US$ 41.4 billion in 2007, suffered a sharp setback after the global slowdown with growth slowing down from 30.5 percent in 2007 to 7.4 percent in 2008. While initial trends show that Indian imports from the US bore the brunt of the recession, recent numbers show that Indian exports have dropped more sharply than imports indicating some resonance in domestic demand (The Financial Express Daily, 2011). Numbers show that the growth of Indian exports to the US slowed down from 10.1 percent in 2007 to 8.4 percent in 2008, while that of Indian imports dropped from 74.3 percent to 6.1 percent.

India's trade has rebounded in 2010 quite strongly, after facing the global financial crisis during the years 2008 and 2009. Over time, the composition of Indian exports has shifted dramatically away from the developed world toward its neighbours in developing Asia. The US share of Indian exports was cut in half—falling from 22.8 percent in 1999 to 11 percent in 2009—whereas the share going to Europe slipped from 27.6 percent in 1998 to 20.9 percent in 2009. [1] Trade and business ties between China and India have increased considerably in the past decade. Mutual trade has increased from around US$ 5 billion in 2002 to more than US$ 60 billion in 2010. The swiftly expanding trade relationship between the two countries could help encourage a mutual interest in regional stability. Coming back more specifically to the trade relationship of India with USA, India has been enjoying a favourable trade balance.

However, consistent trade surplus has two problems. One will be the continuous pressure of the country's currency to appreciate which hurts future export realisation and which makes imports cheaper and thereby hurts domestic production. Secondly, trade surplus means that the country earns more foreign exchange than it needs and therefore build up forex reserves. Earning foreign exchange by exporting is beneficial only when such earning are fully used to import goods of better quality or lower price or import good the country needs but is not capable of producing or buying new technology from foreigners. Let us now look at the composition of India's trade with USA (Table 7.6).

From the table it is clear that the export profile has shifted towards more value added goods and engineering products. Import composition has not seen much change over the period between 2000 and 2008. Out of total imports, only less than 20 percent constitutes those goods, which are capable of value addition in the country. As mentioned earlier, India should readdress its composition of goods that are trading with USA. Theoretically, it is not advisable to have long standing trade surplus with the largest economy. By keeping the trade surplus with USA, we may loose the chance of having more capacity addition in the domestic economy.

Table 7.7 depicts clearly the picture of composition of India and US trade with China. United States exports more electrical machinery, power generation equipments, oil seeds, fruits, plastics etc to China. But India exports more of cotton yarn, fabrics, copper, jewellery, plastics etc. to China.

Table 7.6: Composition of India's Trade with USA

	2000	2008
Exports	Cut and polished non-industrial diamonds, jewellery, textiles and clothing, carpets, shrimp and other marine products. These items constituted over 70 percent of India's export to US	Diamonds and precious stones (22 percent), textiles (22.4 percent), pharmaceutical products (6.7 percent), organic chemicals (5.6 percent), machinery (5.4 percent) and electrical machinery (4.8 percent)
Imports	Machinery, fertilizers, aircraft and aeronautical equipments, medical equipments and organic chemicals	Aviation and aircraft (20.8 percent), engineering goods and machinery (19.8 percent), precious stones and metals (11.9 percent), organic chemicals (5.6 percent), and optical instruments and equipment (4.8 percent)

Table 7.7: Composition of India's and USA's Trade with China

	USA	India
Exports to China	Electrical machinery and power generation equipments, oil seeds, oleaginous fruits, plastics	Cotton yarn and fabric, copper, jewellery, plastics
Imports from China	Toys and games, apparel and furniture	Electrical Machinery and mechanical equipments, mechanical appliances, organic chemicals and mineral fuels and products.

When we look at the import composition of both the countries from China, that US imports toys, games, apparel, furniture etc. and India imports electrical machinery, mechanical equipments, organic chemicals, mechanical appliances, mineral fuels etc from China. This example clearly shows that, there is more space available for trade engagements between India and USA. By engaging more with India in the form of trade and investment relations, United States can get rid of their huge trade deficit with China and thereby it will lead to the global economic balance.

Investment Relations

India has been ranked at the second place in global foreign direct investments in 2010 and is placed among the top five attractive destinations for international investors during 2010-12, according to United Nations Conference on Trade and Development (UNCTAD) in a report on world investment prospects titled, "World Investment Prospects Survey, 2009-2012". India was the 4th most attractive foreign direct investment (FDI) destination in 2010, according to Ernst and Young's 2010 European Attractiveness Survey. The United States is India's largest investment partner, with a 13 percent share. India's total inflow of US direct investment was estimated at more than US$ 16 billion in 2008.

However, during earlier part of the year 2011, Mauritius has led investors into India with US$ 6,637 million worth of FDI comprising 42 percent of the total FDI equity inflows into the country. This increase is mainly because that the multinational companies are using the favourable situation in Mauritius because of the existence of the special treaty between India and Mauritius. The FDI equity inflows from Mauritius is followed by Singapore at US$ 1,641 million and the US with US$ 1,120 million, according to data released by DIPP.

During 2002 to 2005, US investment in China dropped by over 43 percent. In contrast, the US investment in India zoomed 68 percent. But in 2005, the US investments in china totalled US$ 3.06 billion compared to US$ 0.47 billion in

India. Two-third of the US investment in China is in manufacturing sectors. In 2004, transport and electrical equipment accounted for 44.6 percent of the US investments in China. Ironically, these investments boomeranged on the US. Foreign firms in china including American ones became exporters to the US, leading to a yawning trade deficit and raising fears of unemployment in the US.

In contrast, the bulk of the US investment in India flowed into computer software and related areas, followed by consultancy/service sectors. The advantage that India has over China is that a better quality human resource, especially at the managerial levels. India has also a large working age population. India is expected to have 220 million more workers than China by 2050 (Majumder, 2006). India heralds a big potential for investment in retail market. The potential of China has saturated with more than 40 MNC's already in retail business.

India's sizable population and large and growing middle class make it a potentially large market for US goods and services. But the major impediments to expanding the trade and investments into India for US are the, constraints in FDI in many sectors, poor infrastructure, lengthy bureaucratic procedures, poor enforcement of dispute resolution mechanisms, unfavourable labour and other policies, tariff and non tariff barriers to trade, meagre enforcement and inadequate laws in Intellectual property rights and corruption. Maintaining of high tariff rates on imports, India also assesses high surcharges and taxes on a variety of imports. Major non-tariff barriers include sanitary and phyto-sanitary restrictions, import licence, regulations, discriminatory government procurement practises, export subsidies etc are still prevalent the economy.

Almost 2 million people of Indian origin are living in USA. Their participation in the Indian economy has been largely confined to remitting money or investing in equity and bank deposits rather than into more productive investments. Interestingly, research has shown that immigrants were key founders in one of every four US engineering and technology firms, with overseas Indians involved in 26 percent of these.

With the successful launch of the moon mission Chandrayaan, India has catapulted itself into the knowledge domain. It can now be viewed as a key partner in global efforts in expanding the frontiers of knowledge. Chandrayaan included two NASA instruments. This can set the stage for significant cooperation in remote knowledge cooperation, especially in areas such as healthcare and education through US-India Knowledge Initiative on Agriculture Education, Research, Service and Commercial Linkages, established in 2005. The large population of overseas Indians in USA is a repository of knowledge and resources, and can serve as an effective bridge to doing business with India.

The US can play a role in facilitating greater economic exchanges between India and its neighbours by helping develop gateways of trade and infrastructure. American companies need to explore India as a manufacturing hub for export goods, and invest accordingly. Considering that India, a country with a GDP of around US$ 1 trillion, accounts for a mere 1.3 percent of US trade, there is substantial potential to increase bilateral trade.

End Note

1. Calculated based on Direction of Trade Statistics Online Database, IMF, from Prabir De, Trade facilitation in India: An Analysis of Trade Processes and Procedures, working paper No. 95, Asia Pacific Research and Training Network on Trade, 2011.

References

Confederation of Indian Industry (2009): "Indo-US Economic Relations: The Next Decade".

Gargan, Edward A (1992): *"A Revolution Transforms India: Socialism's Out, Free Market in"*, New York Times, March 29, 1992.

http:// www.meaindia.nic.in.

http://newdelhi.usembassy.gov/econ-trade-agri.html.

http://static.globaltrade.net/files/pdf/20100408065304.pdf.

http://www.uscc.gov/annual_report/2007/report_to_congress.pdf.

Krishna, Pravin and Guru Sethupathy (2011): "*Trade and Inequality in India*", NBER Working Paper Series, National Bureau of Economic Research, Cambridge MA.

Lum, Tomas and Dick K. Nanto (2007): *"China's Trade with the United States and the World"*, CSR Report of Congress, Congressional Research Service, Government of USA, January 4th.

Majumder, S. (2006): *"India-US Economic Relations on a High"*, The Hindu Business Line, Financial daily, June 1, *Multinationals"* ISID Working Paper 2007/10.

Panagariya, Aravind (2004): "India in the 1980's and 1990's: A Triumph of Reforms", *IMF Working Paper,* International Monetary Fund.

Pillania, Rajesh K. (2010): *'Indo-China Trade: Trends, Composition and Future'* Journal of Applied Economic Sciences, Vol. 2(12), pp. 129-137.

Talbot, Strobe (2004): *"Engaging India: Diplomacy, Democracy and the Bomb"*, Brookings Institution Press, 2004.

The Library of Congress (2004): *"CRS Report for Congress"*, Government of USA, February 25, 2004.

8

Obama Administration and US-Pakistan Relations

M.J. Vinod

Ties between Washington and Islamabad have been distinguished by a long history of trust and mistrust. It has been difficult to say anything with 'finality' as an 'element of uncertainty and elusiveness' has always existed in the relationship. Pakistan has constantly remained in the eye of the storm due to what a former Pakistan Ambassador to the US called the "tyranny of geography". Besides, Pakistan is situated in a very volatile part of the world which leads Pakistan to be a critical player in both regional and global security (Lodhi, 2009). The US approach to Pakistan has also been heavily influenced by 'personal relationships' at the top. The relationship has also been highly 'militarised' and 'centralised'.

US-Pakistan ties have entered a new phase since 9/11, primarily driven by the US response to the 'war on terror'. Since then Islamabad has assumed a strategically vital yet controversial role in Washington's foreign policy calculations to combat terrorism. Since 9/11, the US government has declared that its larger strategic policy objectives toward Pakistan are grounded on the following goals:

1. Global war on terrorism;
2. Non-proliferation of weapons of mass destruction;
3. Managing Pakistan-Afghanistan ties;
4. Maintaining peaceful relations between India and Pakistan; and
5. Strengthening the democratic fabric in Pakistan.

However, given Pakistan's track record so far, it is bound

to pose one of the 'greatest foreign policy challenges' to the Obama administration. It may be recalled that during his candidacy, Obama had highlighted the following issues:

1. That there was the need to 'shift the strategic focus from Iraq to Afghanistan'.
2. That the Iraqi adventure was a 'strategic blunder'.
3. That Afghanistan-Pakistan-India are interconnected parts of the 'same strategic puzzle'.

The US has increasingly been using the phrase 'smart power' which is a combination of hard and soft power. Moreover, Washington visualizes the need to use "the full range of tools at its disposal, viz. diplomatic, economic, military, legal, political, and cultural. However, there are serious limitations to the extent to which it can exercise leverage over Pakistan. The dilemma in the US approach is that:

1. Traditionally Washington adopted a 'politically soft approach to Pakistan's military and ISI;
2. A 'militarily hard approach towards Afghanistan; and
3. A 'politically' confrontationist approach towards Iran—a country that could provide a viable supply route to the war zone.

Hence, there is the need for a fundamental change in US policy towards Pakistan. The Bush administration it may be recalled had tried to de-hyphenate India-Pakistan ties. The rising economic profile of India also helped the Bush administration to look at India as a separate, sovereign power and not a conjoined twin of Pakistan.

Pakistan's War on Terror and Challenges before Obama Administration

The Obama administration's policies and strategies towards Pakistan have been caught in a web of dilemmas, contradictions and challenges. The following are some of the major challenges before the Obama administration vis-à-vis Pakistan:

1. To convince Islamabad that the real danger to Pakistan emanates within i.e., from the internal extremists and terrorists, and definitely not from India.
2. That certain segment of the Pakistani army and the Inter-Services Intelligence (ISI) perceive the militant groups as a *'hedge'* to be used against India in Afghanistan and Kashmir.
3. That the training Pakistan's military needs to fight insurgents are quite different from the training required to fight a ground war against India. Presently, a large number of military advisers are working in Pakistan to train Pakistani army and paramilitary forces to fight insurgents.
4. That there is need to reduce the 'trust gap' between Washington and Islamabad. It may be recalled that when late Richard Holbrooke, was asked on the 'News Hour with Jim Lehrer' whether Pakistan's security forces were committed to rooting out terrorists, he replied bluntly— "I've rarely seen in my years in Washington an issue which is so hotly disputed internally by experts and intelligence officials".

Admiral Mike Mullen, Chairman of the Joint Chiefs of Staff recognizes that Pakistani military officers also have some grounds for 'mistrusting' Americans. According to him, the cut-off of U.S. military aid in 1990 *led to* a whole generation of Pakistani military officers who have had no contact with Americans, and who suspect American motives. Hence, Mike Mullen made numerous visits to Pakistan to confer with top military officials, as had Gen. David Petraeus, head of Central Command. Pakistan's army chief, Gen. Kayani, and the head of ISI, Lt. Gen. Shuja Pasha, who were regularly conferring with U.S. officials in Washington.

Gen. David Petraeus of the Central Command said on April 1, 2009 that Islamist insurgents pose a growing threat not only to Afghanistan but to Pakistan's "very existence". Bruce Riedel in his book 'The Search for Al-Qaeda' observes that former president Gen. Musharraf had allowed Al-Qaeda to 'regroup' in 'Pakistan's tribal lands'. Moreover, Musharraf and his ISI stooges was 'more an enabler' 'than a foe' of terrorist state formation in Waziristan. Now Pakistan seems to be paying a heavy price for all its follies.

On the one hand the Pakistani's have been nervous about U.S. predator drone strikes against Al-Qaeda and Taliban targets which not only cause collateral damage but also affect public opinion. On the other, Islamabad complains that the US isn't providing the equipment they need. Hence, Islamabad has been demanding cobra helicopters, night-vision equipment, jamming equipment to block the radio broadcasts of the Taliban commander in Swat. They would also like 'small surveillance drones'. On the whole, the 'surgical military strikes' of the U.S. Air Force will continue to be an issue of contention in U.S.-Pakistan relations.

Afpak: Issues and Challenges before Obama Administration

It may be recalled that late Richard Holbrooke had invited the Defence and Foreign ministers of Afghanistan and Pakistan to Washington to give him inputs into Obama's strategic review of the 'Af-Pak' related issues. A large number of additional troops headed for Afghanistan since 2009. Many of these troops have been deployed in provinces close to Kabul which currently do not have a significant international presence and are largely under insurgent control. The challenge for Washington is to hold onto the territory it has gained in Afghanistan and initiate reconstruction projects so that the local population can see the benefits of their presence.

A concern expressed by western officials has pertained to the propensity on the part of some elements of the Pakistani military of supporting militants based on Pakistani soil who cross into Afghanistan and launch attacks. This has long been a source of tension between the two countries, but in recent years some militants allied with Al-Qaeda have also launched a war against Pakistan. The result is that Islamabad follows a dual policy of on the one hand, cracking down on such militants arrayed against the Pakistani state—which includes Al-Qaeda; and on the other covertly tolerating/backing militants who restrict their fighting to Afghanistan. Perhaps Islamabad is now paying a heavy price for this dual policy.

The NATO countries involved in Afghanistan—who make up roughly 55,000 of the 70,000 troops in the country—have differing rules of operation. Some refuse to deploy to combat-intensive zones and others have already set withdrawal dates. Both Pakistan and the United States must both "do more" to fight terrorism. Alternative strategies also need to be adopted. The battle cannot be won by military force alone. The response has to be multi-pronged which requires an approach that encompasses the political, economic and social as well.

To succeed, this initiative must be fought and won on either side of the Pakistani-Afghan border. The US needs to stay focused on Afghanistan and put more pressure on Islamabad." Bill O'Reilly contends that, *"there's no winning the Taliban war unless Pakistan cracks down on the guys that are in Pakistan"*. Success in Afghanistan depends on success in Pakistan and vice versa.

Many of the tribes and developmental challenges are the same on both sides of the border. Af-Pak will remain as long as the US does not move away from its failed policies.

Washington's Aid to Pakistan: Issues and Concerns

A large part of the US aid to Pakistan has gone to the Pakistan army in the hope that Islamabad will cooperate in the US efforts in combating terrorism. The US has pledged huge military/financial aid to Pakistan, though this is conditional on Pakistan's cooperation on the war against terror. Between fiscal years 2002 and 2008, the US has spent more than US$ 11.2 billion. Of this amount, almost US$ 5.9 billion was from the 'Coalition Support Fund' (CSF) programme. This is strictly meant to reimburse countries for the costs incurred in the support on the global war on terror.

The funds were never intended to build 'capacity' within

the armed forces of Pakistan. Washington generally pursues what Christine Fair calls 'supply-driven' aid to Pakistan. In the process large sums of money is pumped into Pakistan's coffers with 'few meaningful accountability requirements. Pakistan has resisted such scrutiny, asserting that once such monies go into Pakistan's exchequer, they become sovereign funds (Fair, 2009: 150-51). Coalition Support Funds account for the majority of the US assistance to Pakistan. Among all the countries receiving such funds, perhaps Pakistan is the largest recipient.

Officially the fund is reimbursement for food, clothing, fuel, ammunition, billeting and medical expenses. Islamabad has been repeatedly submitting receipts to the US Central Command. Pakistan is also one of four countries along with—Israel, Egypt and Jordan—to receive budget support from the US. The purpose of this direct cash transfer to Pakistan is to help it pay off its debt, so that it can spend more on the social sector. Needless to say that there is little accountability in the manner in which the money is being spent (Cohen and Chollet, 2007).

Most of Washington's US$ 11 billion in aid to Islamabad, has gone to the Pakistani Army, in the hope that the then military-led government of former President Pervez Musharraf would fully cooperate with US efforts to prevent the Taliban and other radical groups from returning to Afghanistan. But not only have the Taliban and its allies made a strong comeback in Afghanistan over the past two years—in part due to the safe havens they have enjoyed in the tribal regions on the Pakistani side of the border—but their brand of radicalism has spread outward from the Federally Administered Tribal Areas (FATA), where Al-Qaeda's leadership is believed to be based, into the North-West Frontier Province (NWFP) and beyond even into Punjab and other parts of Pakistan's heartland (Lobe, 2009).

The Obama Administration, has increased 'non-military aid' to Pakistan. Islamabad needs to be held accountable for security in the border region with Afghanistan. It may be recalled that Joe Biden had in his capacity as Chairman of the Senate Foreign Relations Committee introduced a legislation proposing to triple non-military aid to Pakistan. The legislation had authorised US$ 7.5 billion over five years mainly for building schools, roads and clinics. The US moved ahead with the following two legislations:

1. The 'Biden-Lugar' legislation (the "Enhanced Partnership with Pakistan Act of 2008"); and
2. The "Reconstruction Opportunity Zones (ROZ) legislation—which would allow for duty-free access to US markets for certain type of goods produced.

As a presidential candidate Barack Obama had advocated putting more pressure on Pakistan as he feared that Islamabad is preparing for a war against India by using US military aid given to them for fighting terrorists. Asked how Islamabad could be pressurized, Obama said: "we are providing them military aid without having enough strings attached; they are using the military aid – for preparing for war against India" (IANS, 2008).

Washington's ties with Islamabad have grown tense. Pakistan has been unhappy about the pace of deliveries of U.S. military equipment and over congressional attempts to impose restrictions on its supply and use. "We are fighting this war today", a senior Pakistani military official said while describing U.S. assistance as slow and stingy. "What good is it two years from now?" That official and others said there have been long delays in the delivery of helicopters, night-vision equipment and other supplies requested for the army's ongoing offensive against Pakistan-based insurgents.

In recent interviews, civil and military officials in Pakistan drew a sharp contrast between the billions of dollars in assistance that George W. Bush's administration gave, with few strings attached, to then President Pervez Musharraf—a general who came to power in a military coup—and what they see as efforts to condition assistance to the democratically elected government of President Asif Ali Zardari (*The Nation*, 2009). The Obama Administration had requested for US$ 2.5 billion in direct security assistance funds for Pakistan in 2010, a 25 percent increase over 2009.

A US Defence official is on record to have said that Gen. David H. Petraeus, head of U.S. Central Command, "personally gets a daily update—daily, mind you—on supplies shipped to Pakistan. That should give you some sense of how riveted we are on this." However going by the past experiences military aid to Pakistan need to be channelled through 'civilian institutions'. Given the propensity for the aid to be diverted from its original intent, perhaps the time has come for the US to be as transparent as possible about the aid it provides to Pakistan.

US-Pak Ties and Developments in the Swat

The rise of the 'Taliban version of Islam'—which is at variance with the 'subcontinents traditional version of Islam'—is a huge challenge for Pakistan and the region. Until 2007 the SWAT had been a popular tourist destination dubbed the 'Switzerland of Asia'. Till the offensive by the Pakistani authorities, about 75 percent of the valley was under the control of the Pakistani Taliban, which had 'destroyed schools' and 'terrorized the population.' If the authorities in Pakistan have been unable to tackle a home-grown insurgency just hours away from its seat of power, prospects for their cracking down on Taliban and Al-Qaeda forces along the border are grim (Baker and Waraich, 2009).

It may be recalled that just one week after Pakistani President Asif Ali Zardari approved the Swat Valley peace agreement following passage of a parliamentary resolution urging him to do so, the Taliban took over the neighbouring district of Buner in the North-west Frontier Province (NWFP). Subsequently, reactions from the Pakistani public and U.S. pressure, the Pakistan Army deployed paramilitary troops to the region and Chief of Army Staff (CoAS) General Ashfaq

Kayani sent a warning to the militants that the Army would not allow them to "impose their way of life on the civil society of Pakistan". The statement was a positive first step in clarifying Pakistani policy toward the militants. This was followed by aggressive military operations that have ousted the Taliban from the Swat Valley. Since then the U.S. also has stepped up its drone missile strikes in the region, apparently with the support of Pakistani intelligence (Curtis, 2009).

Obama Administration and India-Pakistan Relations

The Obama Administration has been keen to revive the peace process between Islamabad and New Delhi after it was derailed following the terrorist attacks in Mumbai, popularly referred to as 26/11. Taking the relationship forward could be a key factor in helping stabilize Afghanistan. At the same time the Obama administration knows that the US-Pakistani relationship is too important to let the trust gap widen. Obama had argued out his case as follows:

"I will join with our allies in insisting, not simply requesting, that Pakistan crack down on the Taliban, pursue Osama bin Laden and end its relationship with the terrorist groups. At the same time I will encourage dialogue between India and Pakistan to work towards resolving their dispute over Kashmir and between Afghanistan and Pakistan to resolve their differences and develop the Pashtun border region. If Pakistan can look towards the east (India) with confidence, it will be less likely to believe its interests are best achieved through cooperation with the Taliban" (Mohan, 2009: 174).

In the process there is a risk that president Obama may kick-off a 'regional solution for the India-Pakistan impasse. The path to peace between India and Pakistan need not necessarily go through the Kashmir valley. New Delhi needs to be cautious of the US tendency of linking up Afghanistan's security with terrorism in Kashmir. Obama had pointed out the need to resolve the Kashmir problem to reduce nuclear dangers in South Asia. "I will continue support of ongoing Indian and Pakistani efforts to resolve the Kashmir problem in order to address the political roots of the arms race between India and Pakistan". Obama's comments in interviews to *Time* magazine in October and December 2008 and the reported consideration of former President Bill Clinton as a special envoy on Kashmir were at that point of time perceived as indications of a possible pro-interventionist stance. However, President Obama's statement made during his visit to India, perhaps laid to rest some of India's misgivings.

A few years back when Musharraf was at the helm of affairs, a Pakistan scholar Hussain Haqqani surmised the situation as follows: "Fifty eight years after partition, and in the absence of any incentive or compulsion on the part of India to revise the status quo, it might be prudent for Pakistanis to give priority to normalisation and stability in South Asia over settlement of the Kashmir dispute. To make that possible, the rhetoric fed to Pakistanis over Kashmir must end. Musharraf too should try to lower the rhetoric over Kashmir rather than raise his people's expectations in that regard" (Haqqani, 2005).

ISI chief Pasha told the German magazine Der Spiegel in January 2009: "We may be crazy in Pakistan, but we are not completely out of our minds. We know full well that terror is our enemy, not India". These are signs at the top that some Pakistani security officials understand the need for change. But Pakistani defence officials do not perceive the original Taliban, Lashkar-e-Taiba, the Haqqani network and other such groups as enemies of Pakistan. Rather they tend to perceive them as a 'means' to extend their geopolitical reach into Afghanistan and the region as a whole (Innocent, 2009). The implications of such a policy is perhaps self-explanatory, given the internal security challenges Pakistan is going through.

The US must not fall into the trap of pushing for a resumption of official India-Pakistan talks as an end in itself but instead encourage changes in the dynamics of the relationship that will reduce tensions and uproot terrorism from the region. Without concerted action against the perpetrators of the Mumbai attacks it could well embolden groups like LeT to up the ante. The first step is for Pakistan to demonstrate that it will no longer tolerate the export of terrorism from its territory.

As British Prime Minister Gordon Brown has indicated almost three-quarters of all terrorist plots uncovered even in Britain in recent years, had links to Islamic terrorism in Afghanistan or Pakistan. Hence it is often opined of late that the defence of the west begins in the Hindu Kush and Helmand. In other words, as David Miliband, the British Foreign Secretary suggests, "the badlands of Afghanistan and Pakistan are the 'incubator' of choice for international terrorism" (Cohen, 2009). Former CIA Director General Michael Hayden for example believes that the next attack on the US homeland is likely to originate from western Pakistan.

If Pakistan takes decisive action to close down the LeT, India and Pakistan could pick up the threads of the dialogue they conducted from 2004 to 2007. It may be recalled that through this dialogue, the two sides' strengthened mutual confidence by increasing people-to-people exchanges, augmenting annual bilateral trade to over US$ 1 billion, launched several cross-border bus and train services, and liberalised visa regimes to encourage travel between the two countries. There was even some progress on the vexing Kashmir issue with the two governments narrowing the gap in their rhetoric about an ultimate solution to Kashmir. Both sides referenced the idea of keeping the current boundaries intact and making the Line of Control (LoC) that divides Kashmir irrelevant and instituting some type of joint mechanism to facilitate cooperation between the Pakistani and Indian parts of Kashmir (Curtis, 2009). However, since then we have witnessed a downward slide in the relationship.

Observations

Reducing the trust gap between the US and Pakistan

depends on how Pakistan deals with the militants within Pakistan. Perhaps the militants have been briefed that the Taliban is making huge inroads into Afghanistan and that they perceive it as an opportunity to take over a nuclear-armed state. President Barack Obama's decision just days after his inauguration to authorise a new round of strikes by aerial drones within FATA—particularly in Waziristan, suggests that he was unconvinced by Islamabad's repeated promises to curb militants such as JuD.

The fundamentals of US policy towards Pakistan 'have to be reworked'. There is 'no easy solution'. Some prioritizing of issues would be necessary. Long-term goals cannot be sacrificed for short-term gain. US will have to strike a balance between short term measures—such as frequent air strikes against Al-Qaeda and Taliban targets on Pakistani soil; and the long-term goal of enhancing the 'credibility' and 'effectiveness' of Pakistan's 'civilian leadership and institutions'.

According to Gregory Copley, "the WikiLeaks scandal of late 2010, which has revealed to the Pakistan electorate the well-known mutual disregard and cynicism of the Pakistani and US political leaderships, coupled with the death on 14th December of the abrasive US Special Envoy to the region, Richard Holbrooke, will dramatically limit the efficiency of US-Pakistani cooperation—even more than was the case beforeand compound the speed at which strategic change will occur in Pakistan in 2011" (Copley, 2010).

The future direction of the region pivots on Pakistan's ability to overcome multiple socio-economic challenges as well as its willingness to fight terrorism in all its forms within its own borders. Washington will have to coax Pakistan into doing more against the home bred terrorist havens in the tribal areas and in Pakistan occupied Kashmir. Using the terrorists as a means of 'hedging' against India could only aggravate matters for Pakistan. Needless to say, it would further delay the resumption of the India-Pakistan composite dialogue.

Washington's success in Afghanistan will be contingent on fixing the problems and bottlenecks in Pakistan. Pakistan will continue to be the fulcrum of US policy in the region. Three quarters of supplies for US and NATO troops goes through FATA's Khyber Pass. This will be badly affected if the situation deteriorates in Pakistan. While cooperation between the US and Pakistan has increased in leaps and bounds, the 'mutual distrust' continues.

Restoring trust between Washington and Islamabad will be the greatest challenge. President Obama will have to seek 'more flexible and imaginative ways' of managing US political relationship with Pakistan.

References

Baker, Aryn and Omar Waraich (2009): "Pakistan: A Mounting Problem for Obama", *Time World,* Islamabad, Monday, January 26.

Cohen, Craig and Derek Chollet (2007): "When US$ 10 Billion is Not Enough: US Strategy Towards Pakistan", *The Washington Quarterly,* 30(2): 7-19.

Cohen, Roger (2009): "Af-Pak: UK Resolves, America Ponders", *Deccan Chronicle,* October 28.

Copley, Gregory (2010): "Watershed Year for Pakistan", *International and Security Network,* Zurich, 23 December, http://www.isn.ethz.ch/isn/Current-Affairs/ISN-Insights/ Detailing=en&ots627=fce62fe0-528d-4884-9cdf-

Curtis, Lisa (2009): "From Strategy to Implementation: Strengthening U.S.-Pakistan Relations", *The Heritage Foundation,* July 8, http://www.heritage.org/research/asiaandthepacific/tst070809a.cfm.

Fair, C. Christina (2009): "Time for Sober Realism: Renegotiating US Relations with Pakistan", *The Washington Quarterly,* 32 (2): 149-172.

Haqqani, Hussain (2005): 'The Eternal India-Pakistan-US Triangle', *Gulf News,* April 6.

Indo-Asian News Service (IANS) (2008): "US Presidential Election 2008, 'Pakistan Using US Aid for War against India", 5th September.

Innocent, Malou (2009): "Pakistan and the Future of US Policy", *Policy Analysis,* April 13.

Lobe, Jim, "US-Pakistan: Obama-Tied think Tank Calls for a Dramatic shift", http://ipsnews.net/news.aspidnews=44739.

Lodhi, Maleeha (2009): "A Renewed Approach: Finding Stability in US-Pakistan Relations", *Harvard International Review,* http://www.entrepreneur.com/tradejournals/article/200271864.html.

Mohan, C. Raja (2009): "How Obama Can Get South Asia Right", *The Washington Quarterly*, 32:2: 173-189.

The Nation (2009): "Obama Administration Devising Metrics to Measure Progress under Afghanistan-Pakistan strategy", August 30.

9

India-Sri Lanka Free Trade Agreement in Spices

V. Mathew Kurian and A.D. Manikandan

Introduction

Regional trade agreements (RTAs) have made a colossal impact on intra-regional trade, net FDI inflows, employment etc., in North America, Europe and even East Asia while it has been not successful in South Asia until 2000. For instance, intra-regional trade in South Asia has remained stagnated at less than 2 percent in the last two decades (Husain, 2004). However, South Asia has witnessed a considerable change in intra-regional trade since the intervention of India-Sri Lanka Bilateral Free Trade Agreement (ISFTA) became operational by 2000. For instance, the total trade between India and Sri Lanka has been increased to US$ 3,465 million in 2007-08 from US$ 543 million in 1999-00.

Apart from this, the way in which India and Sri Lanka have concluded ISFTA even though there is an enormous difference in size of the economies, attracted a large amount of interest within and outside South Asia.

India is a big country with 80 percent of the aggregate gross domestic product (GDP) of South Asia while it is less than 3 percent for Sri Lanka. Yet, the ISFTA came into operation and credibly working over a decade. This has made a momentous change in perception about regionalism among South Asian countries. This paper has attempted to answer two important questions: (i) Why South Asia needs regionalism?, and (ii) What is the message to South Asia from the impact of ISFTA on spices trade in South India?

Need of Regionalism in South Asia

South Asia is a sub-region of Asia with more than 1.5 billion people living in the eight countries. This is the poorest performing sub region of Asia, and one of the poorest performing regions globally (Kohler and Stirbu, 2007). Though there is a miracle growth with 7.4 percent during the period 2000-2008 but, still South Asia is performing like Sub Saharan Africa, a region with the highest number of people living in extreme poverty, in many areas.

For instance, as per the World Development Indicators (The World Bank, 2010), the access to improved sanitation facilities to population is 33 percent in South Asia while it is 31 percent for Sub Saharan Africa in 2006; prevalence of child malnutrition percentage of children below age of 5 was 41 percent in South Asia while it was 26.5 percent in Sub Saharan Africa during the period 2000-2007; and maternal mortality rate per 100,000 live births was 500 in South Asia, as compared to 900 of Sub Saharan Africa in 2005. It shows that the incidence and intensity of poverty is high in South Asia. But since the inception of globalisation, poverty has been come down, especially in the last five years due to acceleration in the growth in South Asia (Anand and Ghani, 2009).

Globalization

The term 'globalization' was coined only recently. It was first introduced into business literature by Theodore Levitt (1983) in an article published in Harvard Business Review. Since then it has become a familiar word among scholars and even public. Now one cannot come across a 'social discourse' without the mentioning globalization. However, there is no unanimously accepted definition of globalization. There are multiple views. Some consider it as the ushering of a 'New Global Order' and for others it as imperialism under a 'New garb.'

Globalization can be viewed from three perspectives: (i) as a programme or project, (ii) as an ideology; and (iii) as a process. As a programme, it is the creation of transnational corporations (TNCs). These business conglomerates are seeking the maximum freedom of operations across world. The following definition of 'globalization' by Barnevik, President of ABB, an electronics transnational company reveals, "globalization is a freedom for my group to invest where and as long as it wishes, to produce what it wishes, by buying and selling wherever it wishes, and all the while putting up with as little labour laws and social convention constraints as possible" (Gelinas, 2003). TNCs perceive the entire world economy as a single unit. They want the regional and national economies to be integrated with the world economy through neo-liberal programmes, including privatization and liberalisation. 'Market' is being identified as the core vehicle for this integration. So that globalization may be defined as 'marketization' on a global scale.

Globalization also represents an ideology. In this sense, it bears 'neo-liberalism' which promotes a set of coherent beliefs, views and ideas for legitimizing global capitalism. 'Neo-liberalism' is projected as the only one ideology that is legitimate, respectable and possible. The neo-liberal creed may bc summed up in three beliefs: (a) belief in private property, (b) belief in market laws, and (c) beliefs in free enterprises and free trade.

Globalization also may be considered as a process. In this perspective, the world economy has been subjected to a process of capitalist transformation for the last five hundred years. It means that global economy reached the new phase of 'globalization.' The evolution and growth of regionalism is a major phenomenon in this. Some consider regionalism as a 'stumbling' block of multilateralism and others consider it as a 'building' block of multilateralism.

This debate is still continuing among scholars? But one can say that regionalism has made a considerable impact on intra-regional trade in North America, Europe and East Asia that attracted a large scale interest in regionalism among developed and developing countries.

Growth of Regionalism

The growth of regional trade agreements (RTAs) has been one of the major developments in the global trade system in recent years. The RTAs has increased from 50 in 1990 to 250 in 2003 [1] and it is likely to go up to 500 in 2010. Among RTAs, more than 200 are currently in operation across countries and others are in various stages of negotiation. To explain this unprecedented growth of RTAs, researchers have identified a large number of causes which rests on two pillars: (i) the frustration with the present multilateral trading system (Charlz, 2009); and (ii) the US shift from committed multilateralist to ardent regionalist. The World Trade Organization (2002) pointed out that variety of reasons can be attributed to the rapid growth of RTAs.

Most notable among them are as follows: (i) major trading nations in the global economy including, the United States and the EU, are proactively pursuing RTAs as an effective means of attaining trade liberalization; (ii) the successful integration of the NAFTA and the EU substantiated the great benefits that RTAs generate; and (iii) the countries that have not concluded RTAs are gradually encountering economic isolation and consequently increased opportunity costs and are now endeavouring to pursue RTAs. Pant et al. (2009) clearly pointed out that: "today many developing countries fear that regional movements in other parts of the world will adversely impact their trade as regionalism overtakes multilateralism.

The response has been that most of them are trying to get into one regional bloc or the other via regional trade arrangements (RTAs)." In South Asia, as India is a late starter in this regard, the country had only few RTAs in operation, namely the India-Sri Lanka Free Trade Agreement (ISFTA) which came into operation by 2000 and the India-ASEAN FTA which came into operation by 2010 but, many are in various stages of negotiations.

Economists have argued that the frustration with the present multilateral trade system is one of the important reasons of rapid growth of regionalism. Furthermore, it is easier than multilateralism. According to Paul Krugman (1993), regionalism is easier than multilateralism due to four reasons: (i) The large number of participants in multilateral trade negotiations increases the costs of cooperation among the member countries of the WTO/GATT and also complicates negotiations; (ii) Modern trade barriers are more varied, more complex and less transparent than the tariffs and quotas considered in the early GATT Rounds; (iii) The reduced economic dominance of the United States makes the world trade system harder to run; and (iv) Social, political and economic differences between the Atlantic economies and Asia (Japan in particular).

However, a group of economists like Bhagwati and others do not agree with these arguments. They argued that the United States' shift from devoted multilateralist to ardent regionalist is an important reason for the rapid growth of regionalism. Arvind Panagariya and Bergsten have expressed similar views. However, Baldwin (1997) [2] questioned Bhagwati and Krugman's views on the rapid growth of regionalism. He says that "resurgent of regionalism was caused by idiosyncratic events that were multiplied many times over by a domino effect." However, the domino theory can't explain all incidents of regionalism. He says that the "domino theory never tells us why all the triggering incidents of regionalism occurred." It shows that there is no consensus among scholars over the issue of the rapid growth of regionalism in the world.

Regionalism in South Asia

The South Asian Free Trade Agreement (SAFTA), covering an area of consisting of seven countries namely, India, Pakistan, Bangladesh, Bhutan, Sri Lanka, Maldives and Nepal, that came into operation by July 2006. In fact, South

Asian countries have many characteristics which include similarity in culture, ethnicity, geography, democracy etc. that would enable them to go beyond the SAFTA framework and work for deeper integration like customs union and economic union. However, this would be conditional upon working on the constraints including, policy-induced and structural constraints facing with South Asian countries. A rise in problems and challenges faced by South Asian countries, for instance, fluctuations in India-Pakistan relations, is one which restricts effective and credible working of SAFTA.

The sluggish progress of SAFTA is the main reason for a surge in RTAs in the region. In South Asia, India has been the main focus of regionalism due to its growing importance in the South Asia as well as in the globalised world. But India is a late starter in this regard, now the country is too negotiating for bilateral free trade agreements (BFTAs) with many countries within and outside SAARC (South Asian Association for Regional Cooperation). For instance, India and Bhutan, India-Bangladesh, India-Thailand, India-Sri Lanka, India-ASEAN, India-Korea, India-Australia, India-MERCOSUR, India-European Union, India-ANDEAN are few among them.

Besides India, other countries in SAARC viz., Sri Lanka, Pakistan, Nepal, Maldives and Bangladesh are actively seeking and negotiating for RTAs with many developing countries to operate new regional or bilateral trade agreements. The Sri Lanka-Pakistan FTA, Sri Lanka-Bangladesh FTA, Nepal-Bangladesh FTA Nepal-Sri Lanka FTA are few examples of it. Mushroom in regional trade agreements (RTAs) in South Asia that reflect the changes in trade policies and perceptions about regionalism and multilateralism. The rapid growth of regionalism among South Asian countries have exacerbated intra-regional trade and led to the faster integration with the world economy. It is also observed that the degree of economic globalisation has been increased among South Asian countries since the rapid growth of regional trade agreements.

Background of the India-Sri Lanka FTA

India and Sri Lanka have had a long relationship which is reported to be more than 2,500 years old. Both the countries share a legacy of intellectual, cultural, religious and linguistic intercourse over the centuries. In recent years, the relationship has been extended to trade and investment, and strengthened cooperation in the fields of education, culture, defence and also in formulating broad framework on major issues at international level. The ISFTA is one of the important steps which aiming to increase trade and investment. In 1998, India and Sri Lanka signed the ISFTA and the pact became operational by 2001 (Orland, 2008).

The ISFTA is one of the few BFTAs among the South-South Agreements that is working credibly and could be an example for other agreements to follow how bilateral FTA lead the way to a regional cooperation (Kelegama et al., 2007; Joshi, 2010). This is because there is a huge difference between India and Sri Lanka in many areas, for instance, 80 percent of regional GDP is contributed by India while it is 3 percent for Sri Lanka. Yet, India and Sri Lanka concluded and implemented FTA. The most striking result is that the total trade between India and Sri Lanka has gone up considerably since the inception of this FTA. The main feature of this is both countries were benefited. On the one hand, Sri Lanka's export to India has been increased more than 14 times during the period 2000-01-2007-08 largely because of concessions and provisions given to Sri Lanka under this FTA (Yatawara, 2007). [3] On the other hand, India has become Sri Lanka's main source of import (Kelegama, 2006).

Impact of ISFTA on Spices Trade

The impact of ISFTA on trade is one of the best examples of regionalism in South Asia. This is because total trade between India and Sri Lanka has increased to US$ 3465 million in 2007-08 from US$ 543 million in 1999-00. That is 6.8 times increase in total trade during this period. The main features of this can be summarized into two: (i) the net export or net foreign exchange of India has been increased to US$ 2195 million from US$ 455 million during this period, and (ii) the export–import intensity of India has been decreased from 11.28 to 4.45 during this period. This clearly shows that India and Sri Lanka have improved their trade after the inception of this FTA. This is the one of the major outcomes of this trade agreement at aggregate level.

Table 9.1: Trade between India and Sri Lanka, 1996-97 to 2008-09*

Year	Exports (US$ million)	Imports (US$ million)
1996-97	477.41	42.84
1997-98	489.23	30.21
1998-99	437.13	37.68
1999-00	499.27	44.23
2000-01	640.14	45.01
2001-02	630.89	67.38
2002-03	920.98	90.83
2003-04	1319.20	194.74
2004-05	1413.18	378.4
2005-06	2024.67	577.70
2006-07	2258.30	470.33
2007-08	2830.43	634.96
2008-09	2425.92	356.57

Note: *As on September, 2010
Source: Kurian and Manikandan (2011).

Agricultural items have been played crucial roles in trade between both countries until 2000. For instance, the share of rice in export from India to Sri Lanka was 10 percent in 1997-98. This has changed considerably after the inception of ISFTA. Spices export is one of the major export items between India and Sri Lanka in the last 25 years. Therefore, it would be interesting to look at how and to what extend ISFTA has impacted on trade in spices viz., pepper, ginger, saffron, turmeric, nutmeg, cardamom, cloves etc., between India and

Sri Lanka since this FTA came into effect by 2000. In order to analyse data of spices, we have chosen export-import data of spices from 1996-97 to 2007-08 that give comparative picture of pre- and post-ISFTA period.

Pepper is one of the largest agricultural export items to Sri Lanka from India. Pepper export has been steadily increased from India between 1997-98 and 2008-09 with exception of 2004-05 and 2005-06. During 1996-97 India's net foreign exchange from pepper was around US$ 1 million but, it has increased to US$ 16 million in 2008-09 due to the inception of ISFTA.

Despite, some agricultural concessions were made available to Sri Lanka under ISFTA pepper export has been increased to US$ 55 million in 2008-09 from around US$ 9 million in 1996-97. The main feature of this can be summarized into two: (i) India's net export of pepper increased to US$ 22 million in 2008-09 while it was zero in 1997-98; and (ii) import of pepper from Sri Lanka has increased to US$ 16 million in 2008-09 from US$ 2 million in 1997-98. In this contest, a major question is: Who is the real beneficiary of India's higher export of pepper to Sri Lanka?

While analysing disaggregated level data of pepper trade between India and Sri Lanka, it was observed that Kerala is one of the important beneficiaries. This is because of Kerala is the largest producer of pepper and more than 90 percent of India's export of pepper is from the state. But higher production of pepper in Kerala and Karnataka and hike in import of pepper from Sri Lanka in tune with free trade agreement has affected pepper farmers in South India, especially in Kerala during 2002-03. For instance, the average production of pepper in South India was 70,550 tonnes in 2002-03 and 81,320 tonnes in 2004-05.

At the same time import of pepper from Sri Lanka also increased, especially in 2002 and 2003 Sri Lanka exported 6,530 tonnes and 6,160 tonnes which accounted for 9.6 percent and 9 percent of Kerala's total production of pepper (Tudugala et al.). [4] Hence, domestic producers in South India mainly small farmers in Kerala were badly affected due to a sharp fall in price. For instance, the average farm price of pepper declined from ₹ 124 per kg. in 2000-01 to ₹ 60 per kg. in 2004-05 (AMIC, 2011). It is argued that increase in pepper surplus in the world market has aggravated this situation further.

In fact, fall in the average farm price of pepper badly affected small farmers in Wayanad and Idukki districts in Kerala. However, this situation has improved since 2005-06. Latest data (for 2008-09) shows that the total trade of pepper (export and import of pepper between India and Sri Lanka) increased to US$ 55 million in 2008-09, out of which, US$ 22 million is net export. It shows that the ISFTA has played a better role in promoting pepper trade since 2000.

Besides Kerala, Karnataka and Tamil Nadu are largest producers of pepper in South India Latest figures (for 2006-07) show that pepper production in Karnataka has been improved considerably between 2002-03 and 2006-07. This FTA will open new market for farmers in Karnataka and Tamil Nadu along with Kerala as well as threat of higher import of pepper from Sri Lanka.

Unlike pepper, India's export of ginger, saffron, turmeric, thyme leaves, curry and other spices to Sri Lanka is low. But export of these items to Sri Lanka is growing since 2004-05 due to this FTA. The available data in this regard shows that export of ginger, turmeric and other spices (excluding cloves, cardamom, mace and nutmeg) has increased to US$ 3.77 million from US$ 1.59 million between 2004-05 and 2008-09. But import of these items from Sri Lanka to India is found to be zero even since the inception of this FTA. During these years, India's total export of these items has been doubled in the world (Table 9.2). In fact, increase in export of ginger, turmeric and other spices to Sri Lanka and the world would be benefited to all South Indian states, especially Andhra Pradesh and Karnataka, being largest producers of turmeric and ginger.

Apart from this, there is no need for worry to South Indian farmers in the case of competition from Sri Lanka because they are not potential exporters in this regard. By comparison, South India has received the net export or net foreign exchange of US$ 25.5 million, out of which, US$ 22 million from pepper and US$ 3.5 million from ginger, turmeric and other spices in 2008-09 largely because of ISFTA.

Table 9.2: India's Export of Ginger, 2004-05 to 2008-09*

Year	India's Total Exports (US$ million)	India to Sri Lanka (US$ million)
2004-05	81.88	1.59
2005-06	94.04	1.79
2006-07	104.98	2.35
2007-08	138.38	2.74
2008-09	169.70	3.77

Note: Ginger, saffron, turmeric (curcuma), thyme leaves, curry and other spices.
Source: Ministry of Commerce, Government of India.

Like pepper, export of cloves to India is a biggest source of income to Sri Lanka, especially since this FTA came into operation by 2000. For instance, Sri Lanka has received US$ 18 million from export of cloves from India in 2008-09 while it was US$ 2.5 million in 2000-01.

In fact, export of cloves (whole fruit cloves and stems) from Sri Lanka to India was very low in the pre-FTA period. But post-FTA period, India has witnessed an unprecedented growth in import of cloves from Sri Lanka. It shows that Sri Lankan exporters of cloves have been benefited from this FTA. In India demand for cloves is too higher than supply. Therefore, the deficit of cloves is too high in India. This gap has been come down due to higher import of cloves from Sri Lanka. For instance, the net export of Sri Lanka is US$ 16 million from cloves in 2008-09.

In South India, Tamil Nadu is a largest producer of cloves,

although its production is not enough to meet 10 percent of the country. So it can be inferred that an unprecedented import of cloves from Sri Lanka would have an adverse impact in Tamil Nadu. While analysing data of nutmeg, mace and cardamom, it is observed that India's export is very low but, import is relatively high. India's export of these items to Sri Lanka is stagnated even after the inception of BFTA. However, import of cloves from Sri Lanka to India has increased considerably since 2001-02. For instance, India's import of cloves has increased to US$ 5.8 million in 2008-09 from US$ 1 million in 2001-02. As far as South India is concerned, the higher import of nutmeg and cardamom from Sri Lanka to India would have an adverse impact on states like Kerala and Karnataka, being a largest producer of nutmeg and cardamom in South India.

Conclusion

Regionalism in South Asia has been played a crucial role in accelerating the degree of economic globalisation and faster integration with the world economy. The feature of this is increase in volume of trade and benefiting, all. One would be tempted to conclude that free trade agreement could not lead to benefiting all (Stiglitz, 2010). In this background, this paper has investigated the impact of the India-Sri Lanka Free Trade Agreement (ISFTA) on spices trade in South India. While analyzing the impact of this FTA, it was observed that there was an unprecedented growth in trade between two countries since this FTA became operational by February 2000. It is also noted that trade of agricultural commodities viz., spices like pepper, cloves, nutmeg, mace and other spices have been increased considerably.

Moreover, the ISFTA has brought benefits to both countries. In one hand, Indian exporters of pepper, ginger, turmeric and other spices have been benefited. On the other hand, Sri Lankan exporters of cloves, nutmeg, mace and cardamom have also been benefited. It can be inferred that both India and Sri Lanka have got almost equal benefit from spices trade in terms of net foreign exchange. Furthermore, trade in spices has been improved benefits of farmers in South Indian states, especially Karnataka, Kerala and Tamil Nadu since this FTA came into operation by 2000 with sole exception in 2002-03. It shows that importance of ISFTA and could be an example for other South Asian countries to follow how bilateral free trade agreements (BFTAs) lead the way to a regional cooperation irrespective size of the economy.

This gives a lesson to South Asian countries to replicate such an agreement within and outside SAARC. So that South Asian countries should be eliminated the complicated customs procedures at borders and high non tariff barriers (Asian Development Bank, 2006). In sum, the effective inception of regionalism in South Asia based on a larger understanding of the nature of agrarian economy and industry of each partner would enable South Asia to go beyond SAFTA and lead the way to a fruitful regional cooperation like South Asian union.

End Notes

1. Parthapratim Pal has not provided year in this paper.
2. Domino theory is developed by Richard Baldwin in 1997 for explaining causes of regionalism.
3. Yatawara's paper has provided the nature of the ISFTA which include concessions, provisions and commitments of Sri Lanka and India in the agreement.
4. Tudugala and Kularathna have published this paper without mentioning year.

References

Asian Development Bank (2006): "South Asia Economic Report", October, http://www.adb.org/Documents/Reports/South-Asia-Economic-Report/economic-report.pdf.

AMIC (Agricultural Market Intelligence Centre) (2011): "Farm Harvest Prices in Kerala: Growth and Instability", AMIC Info Series-6, http://www.kau.edu/AMIC/Infopercent20Seriespercent206.pdf accessed on 10 March.

Anand, Rahul and Ejaz Ghani (2009): "How will Changes in Globalisation Impact Growth in South Asia", *Policy Research Working Paper (WPS5079),* The World Bank South Asia Region (Economic Policy and Poverty Sector), October.

Atta-Darkua, Charlz (2009): "What Explains the Recent Increase in Regional Trade Agreements? Do you think that this Development will Hamper further Multilateral Trade Liberalisation within the WTO", Department of Economics, University of Essex, 16th January.

Baldwin, Richard (1997): "Causes of Regionalism", *World Economy* 20 (7): 865-888.

Gelinas, J.B. (2003): *Juggernaut Politics-Understanding of Predatory Globalization,* London and New York: Zed Books, p. 3.

Government of India (2010a): *Trade Statistics-India's Trade: Export-Import Data Bank,* Ministry of Commerce and Industry, Department of Commerce, http://commerce.nic.in/eidb/default.asp accessed on September.

Government of India (2010b): *Statistics-Export/Import/Production-Spice/State-Wise,* Spices Board India, Ministry of Commerce and Industry, http://www.indianspices.com/ accessed as on 10 July.

Husain, Ishrat (2004): 'Economic Development in South Asia and the Impact of Globalisation', *Lecture delivered on the occasion of the Middle East and South Asia Conference of KPMG,* Behrain, 11 July, 2011.

Joshi, Vivek (2010): "An Econometric Analysis of India-Sri Lanka Free Trade Agreement, HEID", *Working Paper No. 04/2010,* Geneva, Graduate Institute of International and Development Studies (GIDS).

Kelegama, Saman (2006): "India-Sri Lanka Bilateral FTA: Sri Lankan Perspective on Emerging Issues", http://www.networkideas.org/ideasact/dec09/pdf/Saman_Kelegama.pdf, accessed on 20th July, 2011.

Kelegama, Saman and Indra Nath Mukherji (2007): "India-Sri Lanka Bilateral Free Trade Agreement: Six Years Performance and Beyond", RIS-DP#119, http://www.ris.org.in/images/RIS_images/pdf/dp119_pap.pdf, accessed on 10 March, 2011.

Krugman, Paul (1993): "Regionalism Versus Multilateralism:

Analytic Note", in J. De Melo and Arvind Panagariya (eds.), *New Dimensions in Regional Integration,* Cambridge University Press for CEPR.

Kurian, V.M. and A.D. Manikandan (2011): "Impact of India-Sri Lanka Free Trade Agreement on Pepper Trade in Kerala", *Rajagiri Journal of Social Development*, 2 (2):127-146.

Levitt, Theodore (1983): "The Globalisation of Markets", *Harvard Business Review*, 161 (3): 92-102.

Orland, Brian (2008): "India's Sri Lanka Policy towards Economic Engagement", *Research Papers* Institute of Peace and Conflicts Studies (IPCS), New Delhi, http://www.ipcs.org/pdf_file/issue/1445888596RP16-Brian-SriLanka.pdf, accessed on 10th March, 2010.

Pal, Parthapratim (2011): "Regional Trade Agreements in a Multilateral Trade Regime: An Overview", accessed at, http://www.networkideas.org/feathm/may2004/survey_paper_RTA.pdf.

Pant, Manot and Amit Sadhukhan (2009): "Does Regionalism Hinder Multilateralism: A Case Study of India", *Discussion Paper 09-03*, Centre for International Trade and Development School of International Studies, Jawaharlal Nehru University, India, http://www.jnu.ac.in/Academics/schools/Schoolof International Studies/CITD/Discussionpapers/Regionalism.pdf.

Stiglitz, J.E. (2006): "Social Justice and Global Trade", *Far Eastern Economic Review,* 169 (2): 18-22, March.

Stirbu, Mariana and Gabriele Kohler (2007): "The Economic and Social Environment in South Asia 2006: Implication for Children", *Social Policy Cluster,* UNICEF ROSA, February.

The World Bank (2010): *World Development Indicators 2010*, Washington D.C., USA.

Tudugala, L.R. and N.D.C.D. Kularathna, Report on Export of Pepper from Sri Lanka to India under ISFTA, http://www.srilankabusiness.com/eresearch/pdf_files/ISFTA-Implications/pepperpercent20Internet.pdf.

World Trade Organization (2002): *Compendium of Issues Related to Regional Trade Agreements,* Background Note of the Secretariat, document TN/RL/W/8/Rev.1.

Yatawara, R.A. (2007): "Exploring Sri Lanka's Free Trade Agreements with India and Pakistan: An Exporter's Perspective", *South Asian Economics Journal,* 8 (2):219-247.

10

Future of Civil Society in Pakistan: The Post-Musharraf Era

G. Gopa Kumar

The concept of civil society has always remained a contested terrain in political theory debate. Scholars like Locke, Hegel, Marx, Gramsci, Habermas and others had argued about the significance of civil society through their divergent and critical perspectives. Both critics and admirers have contributed for the strengthening of the concept during the last few centuries. In the contemporary period, with the decline of nation-state sovereignty, the civil society discourse has additional validity and significance. Theoretically, civil society stands midway between family and the State. But it is certainly more than filler in the socio-political process. The issue and relevance of civil society becomes crucial when it comes to the problem of governance in the post-colonial societies especially in the complex era of globalization.

The existence of a very powerful traditional society with competing loyalties of faction, gender, religion, tribe, caste, etc. has further made the role of civil society problematic. With the emergence of liberal democracy and neo-liberal state and technology and communication networks, developing societies had to balance with the uneven forces of both the traditional as well as modern social structures within their complex political systems. As democracy unfolded, these societies witnessed the growth of political parties, media, interest groups, non-governmental organizations, professional associations, moderate religious organizations, human right associations, etc. All these have influenced the political process of their respective countries.

However, the chequered growth of democracy in most of these countries with intervals of military dictatorship and tribal oligarchies had restricted the pace of civil society formation in a healthy manner. Nevertheless, civil society movements were not eradicated altogether. Rather, breakthroughs emerged leading to the arrival of civil society actors and structures. New social movements and their activities are notable in this context. Significantly, the extent of democratic deficit, that is, largely prevalent in the developing societies due to the failure, discredit ability and ineffectiveness of main stream political parties and politico-constitutional structures necessitated the relevance of civil society movements.

For a country like Pakistan, nation with its feeling of liberal nationalism is the most desirable form of civil society. Nevertheless, successive governments in Pakistan have continued to follow a faulty theme towards civil society since independence. Poverty, poor law and order, absence of social justice, corruption, militarization of society, drug trafficking, ethnic violence, religious militancy and sectarianism, corrupted feudal lordism, primordialism, the lack of strong middle class, absence of strong media freedom and above all the constant disruption of democracy constitute the major issues confronting Pakistan.

Consequently, the country witnessed frequent military interventions, meting out tribal justice in civilian process, sustenance of feudal mindset among the civil servants and low credibility for political parties. Attitudes hostile to development are deeply ingrained in the psyche of dominant classes comprising feudal and religious elites. Political parties now function as closed oligarchies and personal fiefdoms preventing openness, transparency and accountability in political parties. [1] Irfan Husain, a noted columnist in Karachi's *Dawn* newspaper, quoted Madeline Albright, former US Secretary of State who called Pakistan as an "international migraine". The country has become a real cause of global concern since it combines numerous negative aspects of nation-building, yet functioning within the frame work of a post-colonial nation-state.

The civil society concept in Pakistan is unique and new although the suppressed democratic tradition can claim a long history. Civil society, as a concept, is characterized by hybrid forms, multiple inheritances and the unresolved struggle between the practices and values of pre-capitalist society and the new modes of social life, between authoritarian legacies and democratic aspirations. It is an umbrella term in Pakistan for a range of non-state and non-market citizen organizations, networks and alliances operating in a broad spectrum of social, economic and cultural field. These include NGOs, trade unions, professional associations, think tanks, pressure groups, traditional, informal faith based organizations, shrines, seminaries, neighbourhood associations, *and jirgas* besides political parties. In addition, many new groups have been created through the development of capitalism. [2] "Civil society is caught up amid the throes of dying order and the birth pangs of a new one. The mid-category responses also bring out the need for more information sharing and dialogue on civil society issues". [3] To be sure, the civil society in the country is still in the stage of infancy.

Towards Experimenting Democracy

The six decades of the history of Pakistan as a nation-state was filled with constant domestic conflict, rise of tribal and religious militancy, regional tensions and related international issues. More than thirty years, the country was under direct military dictatorship and the Constitution was nullified four times. More than the history of being a failed state, it was also a narration of a failed leadership who lacked statesmanship qualities. Even from the beginning, the political process was dominated by domestic pressure groups like bureaucracy, religious leadership and the militant groups. The Kashmir question offered a traditional ground for national unity among military leaders, political parties and religious leaderships. At the same time, the disintegration of East Pakistan in 1971 exposed the limitations of religion as a uniting force across ethnic and regional lines.

Another failure of the leadership was the setbacks it received for aligning with USA and western powers. Not only that Pakistan became a dependent country of USA for security and non-security aspects, but also the country suffered from the ill-effects of cold war. The invasion of Afghanistan by Soviet Union in 1979 opened up another complex history. Ultra-religious forces were raised with the support of Pakistan, to strengthen Taliban in Afghanistan and resist the communist regime there. With the exit of Soviet forces and the subsequent disintegration of Soviet Union, the regional balance of power was upset.

Already by this time, the Northern provinces of Pakistan like NWFP and Baluchistan became vulnerable to intense drug trafficking and clandestine militant activities for supporting Taliban. The rise of Osama Bin Laden and his forces controlling parts of Afghanistan and Federally Administered Territorial Areas of Pakistan had far reaching consequences in the marginalization of democracy in the country. The setback of political parties in offering resistance to religious and medieval rule, besides the endemic corruption among the parties, enabled the military to tie up with corporate forces in USA and control political power in Pakistan.

The political events following September 11, 2001 made a world of difference for Pakistan. The military regime got entrenched in the political system and main stream political leaders were either exiled or arrested. The USA needed full support of the military leadership in Pakistan in repulsing Taliban and Al-Qaeda forces from Afghanistan and possible hide outs in the northern regions of Pakistan. The military in an uncontrolled situation not only cracked down civilian political leadership but also entered into conflict with judiciary and the media. In fact, this was the situation which enabled the spontaneous rise of civil society movements in Pakistan. The

Musharraf regime pressurized the judges for favourable judgments and threatened print and electronic media for highlighting anti-governmental news.

Press censorship was severe and the non-governmental T.V. channels and even the internet system were jammed. Fake elections were conducted for justifying the continuation of General Musharraf as the President. He also advocated parliamentary reforms by changing electoral laws, providing 33 percent reservation for women in Parliament and strengthened local democracy at the village level. In the meantime, ethnic division sparked up between Baluchs and Sindhs, Mohajirs and other sections. Religious revivalism led to the increased role of *Madrasas in all walks of socio-political life*. The grip of religious orthodoxy over the political process and the continued ethnic conflicts across the country shook the edifice of stability presented by the military.

By 2004, civil society in Pakistan was caught between the cross currents of modernity, religious extremism, forces of globalization, extreme deprivations, sub nationalism and limited opportunities. [4] It was doubtful whether the policies of deregulation, privatization, and success of local self government could offer a gradual change for strengthening democracy and civil society. Given the geo-political and strategic significance of Pakistan, USA felt the urgency of brokering peace in Pakistan.

The fake victory of Musharaff in the Presidential elections held in 2007 was challenged in the Supreme Court by the civil society groups. Anticipating an adverse judgment, the President sacked Chief Justice Iftikhar Choudhary and imposed an emergency in the country. The civil society groups strongly protested against these undemocratic steps of Musharraf. The compromise was offered at the initiative of USA by bringing together Pakistan People's Party leader Benazir Bhutto and President Musharraf for a political tie up and marginalizing the role of Pakistan Muslim League (N) and its leader Nawaz Sharif. The electoral battle for Parliament began to surface in this background with the visits of Ms. Bhutto and Mr. Nawaz Sharif. The latter was deported by the regime. On the eve of elections, the democratic apparatus broadened further and all the major political parties were permitted to campaign with the people. President Musharraf and his supporting organization Pakistan Muslim League (Q) still hoped to win a mandate with the people.

President Musharraf had nurtured religious militancy in the past but by 2007 it has became a powerful force against him. The pro-American policies of military regime estranged him from the ethnic and militant religious groups. The military crack down on *Lal Masjid* (Red Mosque) in Islamabad in July 2007 invited more troubles for the government since the network of religious order hostile to US-Musharraf axis became widespread. Hence, it was not surprising that Benazir was attacked on 18 October 2007 on the same day of her return to Pakistan after years in exile.

Many believed that Benazir's decision to forge an agreement with military dictator would not ensure a smooth transition to democracy. This was because a fresh balance of power was to be established between military and civilian institutions. Again, the negative forces are too strong to allow any political player to establish legitimacy. Benazir's tragic death on December 27, 2007 exposed the limitations of institutional strength and simultaneously demands a new responsibility for civil society in Pakistan. [5] As a matter of fact, Musharaff's authoritarian style had given Al-Qaeda a recruiting tool for militancy as well as divide Pakistan's secular and moderate sections. The Jihadis want Pakistan to be an Islamic emirate and found Benazir in the opposite side. [6] Although we could notice that the mainstream parties were all right wingers, the PPP carried a special identity of being more moderate force.

With the exit of Ms. Bhutto the moderate constituency has weakened further. The civil society movement which campaigned to have the Chief Justice Choudhary restore his job faced an immediate set back in this situation. The American plan to co-opt Benazir with the Musharraf regime on a power sharing arrangement also received a great set back in this context. Indeed, the Musharraf regime was also weakened by the death of Bhutto. The national situation became very grim. Arif Rafiq notes that 'the nightmare scenario envisioned by many in Pakistan—a nuclear armed country actively targeted by Al-Qaeda and Taliban—could become a reality. [7]

He feared that terrorist who feed off of political instability will continue to gain, while the country's poor and illiterate majority will worsen to loose ground. On the whole, the general feeling in the society was that—(i) Bhutto's death was a major set back to liberal and moderate politics, (ii) the incident underlines the continued threat of extremism to the Pakistani state and society, (iii) the prospects of reviving democracy in Pakistan has also reached a crucial point. [8] According to Ayesha Siddique, political fundamentalism adopted by the global elite (military industrial interests) was responsible for the killing of Bhutto.

The global elite worked out, by using her life and death, a 'perpetual disorder in Pakistan for many decades to come so as to keep this strategically located country of 165 million people under full control and use it objectively to get full control over 'Euraisa' because only then they can meet future challenges from China and ECO countries'. [9] According to Ahson Saeed Hasan, 'Pakistan is miles and miles away from the establishment of a civil society. The involvement of toxic Islam has only polluted the environment and has taken the nation away from any prospects of prosperity and a healthy future for the generations to come. There seems to be no way out. The monster of fundamentalism has left the country and its people paralyzed and doomed.' [10]

The post-election scenario witnessed a major change towards strengthening democracy after eight years of military dictatorship. Although the election results produced a hung parliament, the PPP, the biggest winner, and the PML (N), the second major winner, resolved to iron out their differences and work towards conciliation. The coalition government started with a positive note and in the first session of the new House, it expressed the hope that the sacked judges of the Supreme Court will be restoring their status. The *Muree Declaration* of March 30, 2008 further consolidated this position.

The parties reaffirmed the status of 1972 Constitution and made it clear that as in every parliamentary democracy, the President is bound to function according to the advice tendered by the Prime Minister. The political signals were very clear—President Musharraf could resign before the impeachment take place in the Parliament. The *Muree Declaration* offered thirty days deadline for the new government to restore the judiciary of the status of the pre-emergency period. The assurance could not be met by Prime Minister Yusuf Raza Gillani so far because of the second thoughts from the PPP Acting Chairman Asif Ali Zardari (who was alleged to be pressurized by President Musharraf of opening his corruption cases and charges). The situation did not produce a stalemate since Nawaz Sharif was willing for giving an extension for the deadline of April 30 to a sooner date.

There was some improvement in the domestic scene. The newly formed government in the Pakistan, at the centre as well as the North West Frontier Province resorted to distinguish between Al-Qaeda recruits and the local insurgents. The latter need to be drawn into the mainstream while the former must be dealt with severely. The policy yielded few results and a militant leader in South Waziristan, Baitullah Mehsud, ordered his men to cease fighting. [11] Again, the news on the attempt by terrorists to kill Afghan President Hamid Karzai on April 27, 2008 at Kabul was received with shock from Pakistan side. The Afghanistan's intelligence Chief Amrullah Salah stated that the plot was hatched in Bajaur and North Waziristan tribal areas and Peshawar. [12] Pakistan rejected these allegations but nobody can assure that militant activities have declined in northern tribal areas.

The marginalization of military power in Pakistan may be a temporary development. President Musharraf's exit in August 2008 cannot be construed as a real exit of military power structure in Pakistan. Similar power centre can arise in future as well. Actually, "there is no history of Pakistani military dictators leaving on their own. If there is anything 'new' about Pakistan today, it is not that democracy has returned but that a dictator agreed to leave without resistance". [13]

At the moment, the civilian government's ability to control ISI activities in clandestinely promoting cross-border terrorism and militancy is questionable. An unholy nexus between militants and ISI had allegedly emerged in vulnerable pockets of North West and this could challenge the credibility of Gillani government, both at the domestic and international front. In the meanwhile, the PPP Acting Chairman, Asif Ali Zardari, who won the presidential race in September 2008, did not show any inclination to restore the judges in their original position allegedly because of the fear of the revival of the cases against him. This became another issue in the path of constructing a new civil society in the post-Musharraf era.

Traditional forces and mainstream actors are still powerful in deciding the destiny of the country. Again, with the exit of Musharraf from power, the two major parties missed a rallying point in strengthening this nascent democracy and the new developments also add to confusion. While President Zardari declared that Kashmiri militants are terrorists and Pakistan will not support such attempts, it was quickly challenged by Opposition Leader Nawaz Sharif. The statement from the President (which he later tried to revise) and the reaction from the opposing parties indeed expose the substantial differences among liberal forces in the political system. The civil society's response to this controversy and the American attacks on Wazirstan to flush out the militants was also not reported to be very strong. With the Mumbai blasts occurred during December 26-28, 2008, the role of the failed Pakistani State once again came to the focus of the global community.

Altogether, four distinct centres of power in Pakistan could be noted at this juncture—the democratically elected civil government, the military, the Inter Service Intelligence and the jihad elements. The convergence and divergence of any of the two actors or more bring problems to the civil society. The space of civil society is therefore shrinking and the UN itself is in a real dilemma to persuade Pakistan from discouraging the extremist elements. While Pakistan herself is becoming a victim of the poisonous medicine she had prescribed to others, the uncontrolled growth of religious extremism, in any case, is raising huge problems for humanity in the 21st century. Critics point out that the post-Mumbai blasts scenario was thus cleverly manipulated by the ISI, military and the extremist elements to promote strong anti-India sentiments and conversely pushed the adolescent civil society into a weak defensive position.

The return of former Supreme Court Chief Justice Iftikhar Choudhary to the post on March 25, 2009 was another unique development for which the civil society, however, cannot take much credit. With the increasing pressure on Zardari's government both from the military and USA mounting up besides the agitational politics led by Nawaz Sharif, supported by opposition parties and civil society groups, the beleaguered President had no other option other than bringing back the deposed Chief Justice to his office. It opened up new space for civil society activism with the media highlighting the cause of freedom and human rights.

The plight of religious minorities was exposed further not only by the operation of blasphemy law but the manner in which saner voices were submerged. In an attempt to save an illiterate Christian woman, Aasia Bibi, Salman Taseer, Governor of Punjab, faced 26 bullets from his security guards in 2011. This assassination 'shook liberal and secular Pakistan to the core', argues Pervez Hoodbhoy. While Assia Bibi continues in solitary confinement in jail, Justice Pervez Ali

Shah, the brave judge who sentenced Taseer's murderer, fled the country. [14] The condition of the society and political system itself can be evaluated though these developments.

The US ability to eliminate Osama bin Laden on May 1, 2011 from Abotabad by keeping Pakistani security forces in dark surprised many observers'. Subsequent developments had shown that the traditional alliance between America and Pakistan is cracking and the US is no more willing to trust Pakistan for fighting against terrorism. The recent developments had shown that both the groups (the conservative religious groups and the civil society groups) are vying each other in the truncated space of Pakistani nation-state. The dilemma of Gilani's government in facing the pressure from military as well as from judiciary for separate charges was another problem that confronted an elected government which could not deliver good governance. The US exit from Afghanistan also pushed the Pakistani government into further predicament on security and internal tension.

The Musharraf regime was culpable for enacting the National Reconciliation Ordinance in 2007 which was nothing but an opportunistic attempt by the military general turned civilian president to negotiate with the Pakistan People's Party. The law exempted more than 8,000 culprits including Asif Ali Zardari from their criminal charges. But the judiciary did not consent to this controversial law of granting immunity to politicians with criminal charges. Later, when Zardari became the President and Mushraff left Pakistan, the legal issue re-emerged in the political system. In early 2012, Prime Minister Yousaf Raza Gilani was charged by the Supreme Court of Pakistan for upholding NRO and protecting President Zardari.

This led to another political crisis between judiciary and parliament. The Parliament passed a resolution on Januray16, 2012 asserting its supremacy but the judiciary was not yielding. While judiciary insists for the prevalence of the rule of law, the military is waiting for a setback to the civilian leadership so that it can replace the Gilani's Government.

It is well known that military has scores to settle with the civilian leadership over the *Memo gate scandal (it* was alleged that Prime Minster Gilani tried to seek US help to prevent a military take over in the aftermath of the killing of Osama bin Laden in May 2011). But unlike the past it is not easy to effect a *coup d'tat.* The civil society is more sensitive to anticipate military's game plan although it has not become a major force to reckon with. Nevertheless, the impact of democratic developments in North Africa and West Asia consequent to the social media networking has made the situation somewhat difficult for authoritarian forces.

Future of Civil Society

The future of Pakistan as a modern nation state depends on a variety of factors and forces that operate both in domestic and international scene. Religious militancy and ethnic conflict have clear linkages with the spill over of terrorism from Al-Qaeda and Taliban ranks. Pakistan relationships to India, Afghanistan and Pakistan are crucial in this context. Although the Kashmir issue received a temporary set back in the recent past due to internal problems in Pakistan, Prime Minister Yousaf Raza Gilani tried to dispel the impression that his government will compromise on Kashmir issue.

Again, the success of the proposed 7.4 billion India-Pakistan-Iran gas pipelines depends on improving bilateral relations between India and Pakistan as well as rooting out terrorism in Pakistan. The future of US-Pakistan relations is another matter of grave concern. Under Obama's Presidency, Pakistan did not get the kind of support it earned on the terrorism issue unlike in Bush administration and the killing of Osama Bin laden by the American troops inside Pakistan was a great set back. However, the US cannot bail out on the charge from international community that to a great extent, she was responsible for the present predicament in Pakistan.

It is certainly reassuring to note that the constituency for enlargening the civil society is widening in the post-Benazir scenario. The popular discontent on Musharraf's leadership increased and his ambition to divide the ruling coalition failed. The mass appeal and morale of civilian leadership was very high in contrast to the unpopularity of the former President, especially on the eve of his exit. His stepping down from power promoted, indirectly, strong elements for strengthening democracy in the country.

Any attempt to resist the moves of Gilani's government, by the beleaguered President would have invited sharp reactions from civil society. The election of the new President, Asif Ali Zardari was a smooth affair and his victory was a foregone conclusion. Nevertheless, the lawyers, human rights activists, media workers, students and youth sections had to anxiously wait for long to restore the judiciary with the pre-emergency status and reinstating former Chief Justice Iftikhar Choudhary and his colleagues to their previous position. The mainstream political parties, the Pakistan People's Party, the Pakistan Muslim League (N), and the Awami National Party etc cannot ignore long the expectations raised by the civil society in the post-Musharraf scenario.

If they fail, Pakistan will once again fall into the trap of military leadership or fall into chaos orchestrated by militant groups. The rise of militancy beyond the control of civilian government is a matter of grave concern. All these trends were evinced when Gillani's leadership was charge sheeted by the judiciary on the one side and the threats emerging from military following the *Memo gate scandal*. The survival or dismissal of civilian government is going to be an acid test in democracy. Unlike in the past, military cannot dismiss an elected government for fear of internal and international criticism. There is no guarantee that Supreme Court will ratify military's action and the public is more sensitive to this type of actions.

The civil society in Pakistan strongly reacted against the Mumbai blasts and reminded the Zardari government that it cannot stay in a state of self-denial. The statement issued by human rights activists, women's rights activists, teachers, labour leaders and journalists was very touching "Success neither in the fight against terrorism nor in defending the nation's integrity can be guaranteed by arms alone. The way to end the abuse of belief for politics or for terrorism, there being little difference between the two, is going to be long and hard. The task cannot be accomplished without the whole-hearted support of a fully informed and wide-awake society. The returns on investment in people's food security, education, shelter, health cover and creation of adequately rewarding employment or both men and women will be infinitely higher than on resources expended on guns and explosives. This can be achieved through regional cooperation and trade liberalization". [15]

In this context, it is worth remembering the theory of Samuel Finer who rightly observed that the administrative process in many developing countries revolve 'around a cycle of regimes' beginning with democracy and its experiments, failure of democracy due to internal fighting and corruption of party leaders and other elites inviting interference from army for ensuring stability, the third stage beginning with military coup and take over of the administration, the next stage emerging with public discontent over military dictatorship leading to the launching of the *movement for restoration of democracy* and the final stage culminating with the exit of dictatorship and the return of civilian rule.

However, Finer believed that the political cycle will continue and democracy will be destabilized again in nascent democracies. The history and politics of Pakistan, unfortunately, fits into this framework so far. The potential space for the civil society has to be interpreted in this situation. It can act as a powerful mediator in the political system, negotiating with the divergent actors and forces and ensuring legitimacy, rule of law and preservation of human rights. Over a period of time, proper grounding of democracy could be achieved. In other words, where conventional structures of democracy fail, civil society can emerge with a softening effect. But can civil society deliver governance?

Pakistan had a long history of authoritarian mode of governance and its constitutional record is a poor one. But one matter is clear in each of its constitutional documents. The country was formed for Muslims and was to be guided by Muslim practices. Democracy was implied but it was never made the dominant feature. Its quest for democracy was compromised long ago. Politicians, bureaucrats and soldiers all failed to deliver on their promises. [16] Yet the civil society in Pakistan is willing to give political leadership another chance to strengthen civilian government, devoid of the influence from both military and religious militancy.

Actually, the civil society experience itself is very young compared to the checkered experience of democracy. Gen. Talat Masood remarks that, "the civil society is new phenomenon in Pakistan. It needs to be further strengthened so that, it could put the political parties on the right course. As of now the parties are authoritarian and feudal in character". [17] We could clearly notice the basic deficiency in democracy here. Gen. Ayub Khan, the first military ruler, soon after taking over power in1958, remarked that military will not stay in power for a single minute once the country is saved from internal challenges and anarchy but hastened to add that military will not move out from power for a single minute until the chronic issues of instability and disorder in the political system are properly eliminated.

Did this strange political process change remarkably compared to 1958 period? If at all there are some changes in the situation, it was due to the role played by civil society movements but they are of recent origin. It needs to be sustained but the imponderables are too many.

The ideal option for Pakistan is to modernize her society with an active middle class and harmonize the virtues of Islam to invent a *Pakistani model of democracy*. The bitter lessons should serve as an eye-opener to the political leadership. The return of parliamentary democracy in the country is offering a unique opportunity for advocates of civil society. The media freedom in the country is also on the rise compared to the non-liberal phase in 2000. The SDSA work on *The State of Democracy in South Asia: A Report* had cited Freedom House Index to highlight that media is improving its record in Pakistan recently. [18]

However, if parties fail again in providing development and good governance; the country may fall into chaos (the election to Supreme Court Bar Association also witnessed strong dissent of the lawyers to the attitude of political parties, especially the ruling dispensation, in their failure to address the issue of reinstating judges).

Unlike the past, there are three major challenges in Pakistan's democracy; religious fundamentalists backed by powerful militant groups, the power hungry military leadership, waiting for the right opportunity to strike and seize power, and the corrupt political leadership. The alleged ISI links with terrorists, the recent recruitment of Taliban minded youth in the Pakistan army, the fragility of political process in Afghanistan and the imperialistic attitude of United States in the region are raising big issues for the growth of civil society. These developments also explain the pathological weaknesses of the civil society.

Democratic elements in Pakistan will strengthen only if liberal education overtakes the primacy of religious education imparted by *madrasas.* Civil society need to come forward and focus on educating the youth and women. Even now more than 50 percent of the population is illiterate and survives on a low income. This situation should change for better. Again, it is heartening to note that judiciary is more responsive to constitutionalism and reject both the unhealthy interference from the military as well as corrupt practices from the party system. More than at any time, the country is at a cross road. A stable Pakistan is also strength for its neighbours, especially India. Pakistan must choose between stability and disorder, consensus and conflict, civil society and militancy, democracy and authoritarianism. Whither Pakistan?

End Notes

1. Ijaz Jul, *Pakistan and Civil Society* (June 25, 2004), http://www.chowk.com/articles/7674.
2. Civicus index on civil society, Occasional paper series, Volume 1, Issue II, p.1, http://www.civicus.org/new/media_pakistan.pdf.
3. Ibid., p. 26. http://www.civicus.org/new/media_pakistan.pdf.
4. http://www.chowk.com/articles/7674.
5. http://opendemocracy.net/article/conflicts/india-Pakitan/benazir-bhutto.
6. http://commetistree.guardian.co.uk/paul.cruickshank/2007/12/plan-b-for-Pakistan.html.
7. http://www.socialistworker.co.uk./print_art.php?id=13871/ Arif Rafiq the Pakistan Policy blog http://www.merinews.com/catfull:jsp?articleID.
8. Askari Rizvi, http://www.deence.pk/forums/strategic.geopolitical.issues/8881-Pakistan-after-benazir-bhutto.html.
9. http://www.opendemocracy.net/article/onflicts/india-pakistan/benazir-bhutto
10. http://ourvoicestogether.wordpress.com/2007/12/28/pakistan-life-after-benazir-bhutto
11. "Ominous Warning", Editorial, *The Hindu*, Thiruvananthapuram, May 1, 2008, p.10.
12. Ibid p.14.
13. Hasan Suroor, What's so new about the 'new' Pakistan? *The Hindu*, Thiruvananthapuram, September1, 2008. p.12.
14. Pervez Hoodbhoy, *A Long Sad Year After Salman Taseer's Killing, The Hindu*, January, 4, 2012, p. 13.
15. *The Hindu*, January 5, 2009, p. 10.
16. Lawrence Ziring (2005), *Pakistan: At the Crosscurrent of History,* Manas Publications, New Delhi, pp. 277-287.
17. E-mail reply to the author from Gen. (Rtd.) Talat Masood dated June 27, 2008. See also his article appearing in *The Hindu*, June 26, 2008, p. 11.
18. SDSA, *State of Democracy in South Asia: A Report*, Oxford, New Delhi, 2008, p. 51.

11

Political Economy of Pakistan: Security Implications for South Asia

Rahul Tripathi

Contemporary Pakistan

Contemporary Pakistan has been characterized by a number of formulations which often border on the strategic location where those perceptions come from. According to the US state department, Pakistan has been characterized as one of the most dangerous places on earth today. Closer home, India describes Pakistan as a state that has consciously followed terrorism as an instrument of state policy which now seems to be rebounding on itself. On the close western flank, OIC regards Pakistan as a key strategic and economic partner which will be central to peace and security in the neighbourhood. Even the multilateral economic institutions have their own perceptions and projections about Pakistan which often tally with the lenses with which they would like to see the nation.

What is it that generates much of such projection about Pakistan? First and the foremost, the growing attacks over civil and military targets across Pakistan by radical Islamic groups makes ordinary life of citizens insecure and the ability of the state to deal with them more vulnerable. Second, Pakistani Army, perhaps the only predictable factor in the nation's political economy faces a unique crisis of credibility from within and outside even as it tries to overcome the challenges posed by local insurgencies. Thirdly, with the growing friction with the US after the covert operation that supposedly eliminated Osama bin Laden, any further escalation of war on terror (with increasing drone attacks) the discontent within the moderate as well as extreme elements within the Pakistani society is likely to multiply manifold.

There is a growing questioning of the net gains the nation has had in its complicated association with the US backed strategy in Afghanistan during the last three decades more than ever before. Fourth, the fragility of the ruling dispensation, with Pakistan still to come to terms with its latest democratic experimentation and unpredictability of the political trust quotient within the leading political parties, one is not too sure when Pakistan may go back to its cyclical aberrations.

Finally the economy, which has been a rather neglected area of study by strategic scholars and analysts of Pakistan: while the nation appears to have done fairly reasonably in terms of its macro indicators, GDP growth rate, foreign exchange reserves, inflation, the story at the micro level reflects increasing disparity and a peculiar demographic trend which is likely to leave a growing number of youth with inadequate avenues for engagement a ready fodder for radicalization enterprise. That such as scenario exists against the backdrop of massive dose of foreign civil and military funding till now taken for granted, but now becoming uncertain, makes the situation even more complicated.

One therefore needs to do an exercise of interweaving and interlinking these divergent yet complicatingly related trends in order to assess the challenges that contemporary Pakistan faces. Such an analysis has to include factors which are both extrinsic and intrinsic to Pakistan's larger political economy of development and must not overemphasize on the security obsession that always predominates discussions on Pakistan.

Evolution of the Nation State

Pakistan as a nation was created out of the basic premise of Islam as constituting a nation, which laid the basis of the partition of the South Asian subcontinent. While the premise saw its inherent contradictions with the creation of Bangladesh in 1971, the internal fissures of what still constituted Pakistan remained subject to several tensions from within based on sectarian, provincial and ethnic lines as the country faced several autonomist and separatist movements during the course of its post independence history. With the inability of stable political and constitutional order to emerge and the ensuing instability in this period, the institution of the Military aided by civilian bureaucracy became the core of the centralized state, which repeatedly took upon itself the task of running Pakistan as a stable entity, with there being periods of democratic aberrations and enjoyed relative autonomy from other structures of the society.

As Akmal Hussain points out:

'The military and the bureaucracy at the time of independence acquired a predominant position in the structure of state power. This was due to the fact that the institutions in civil society during the colonial period remained weak...Gradually the army and bureaucracy developed symbiotic relationship. There was also a shift overtime in the social origins of the officer corps...accompanied by politicization of officer corps as they were subjected to the ideological influences from the right and the left wings of the political spectrum. The result was an undermining of the ''relative autonomy'' of the military bureaucratic oligarchy, and its associated difficulty in constituting a credible civilian facade'' (Hussain, 1998).

The stabilizing agenda of the military bureaucratic combine got its further legitimacy with the western donors and friends showing a greater preference for Pakistani Generals rather than its politicians for ensuring that Pakistan fitted well in their larger geopolitical and strategic objectives in the region. It was not surprising therefore that periods of military rule in Pakistan were also incidentally the periods of fairly stable economic growth pumped in by significant inflow of aid as well as investment (Burki, 2007). On the contrary, the periods of democratic aberrations were the ones which were politically unstable and economically insecure to the extent that at times the political economy of the Pakistani state became synonymous with the deepening links that it created with inherent interest of the military bureaucratic combine (Waseem, 1994).

So long as the combine saw itself as a redeemer of the stability of Pakistan state, without trying to change the basic character of the society one saw it as providing perhaps the much needed space to Pakistan to project itself as a viable unit. But as it sought to pursue a well defined agenda of islamization under Zia-ul-Haq, who came to power through a coup in 1977 in one of the most adversarial modes by overthrowing and executing an ex Prime Minister, and the convenient cover which was provided for it by a more than willing anti Soviet sentiment in the West, perhaps was destined to change the course of future history of Pakistan and perhaps of the region in a way not seen earlier.

The Soviet intervention in Afghanistan and Pakistan's use as a proxy by the United States by propping up the mujahideen, enabled Pakistan not only to legitimize the internal policies of islamization and radicalization, but also gave legitimacy to the structures supporting the political economy of the military state. Both were in fact to become reasons for the kind of crisis that Pakistan currently faces as shall be elaborated in the later section of the paper.

The evolution of the bureaucratic-military state interspersed with democratic aberrations was juxtaposed against very complex ethnic and sectarian fault lines at multiple levels that the Pakistani society faced. Even as the state continually attempted to maintain a top down structure for the sake of a united Pakistan, provincialism continued to figure in as recurring feature in some of Pakistan's restive provinces particularly Baluchistan and Sindh. While the former faced as many as three phases of insurgencies, the latest one leading to a full fledged assault against the tribal groups led by Bugtis, the latter saw in the 1970s a movement for autonomy fuelled by a perception of increasing migrant influx and the resulting threat to identity and resources.

Karachi, Pakistan's financial capital in the 1980s and 1990s saw full fledged battles between the security forces and the MQM activists on one hand, and continuing clashes between members of Shia and Sunni community on the other. Meanwhile the perception that the rest of the provinces were being dominated by a military-bureaucratic complex led by Punjabis gained further ground as the inter-provincial economic disparity also deepened as Pakistan embarked on economic reforms in the 1990s. While it would be wrong to say that Pakistan alone was subject to such an economic divide, what made the situation perhaps more complex was that it came against a backdrop where social and political fault lines already existed to make any potential conflagration a dangerous one. And that is precisely what happened when we analyse the nature of the current crisis in Pakistan from a political economy perspective.

Political Dimensions of Pakistan's Economy

Politics and economics often shape and influence each other and this applies all the more to South Asian countries where the agenda of state consolidation and economic development has gone hand in hand. Pakistan, like all the other countries of the neighbourhood has been no exception to this. While there has been an extensive discussion on Pakistan's politics and security and its economic performance separately, there have been limited studies on how the two have been interlinked and intertwined in shaping contemporary Pakistan. It has often been mentioned that phases of military rule in Pakistan have been more conducive for economic growth and stability as compared to those of unstable political coalitions during democratic dispensations (Hussain, 2009).

The present section attempts to locate some such political bases of Pakistan's economy by listing out a set of internal as well as external markers which try to see this linkage in as it affects contemporary Pakistan. Both the markers themselves are not mutually exclusive, but they are at times overlapping.

Still they offer us an avenue to cull out a deeper insight of the contemporary situation.

Internal Markers

The Growth and Development Paradox: Despite its supposedly satisfying levels of economic growth in the recent years, developmental indicators in Pakistan have remained fairly low. Pakistan Ranks in the lower quarter of major human development indicators. Pakistan therefore, perhaps rightly so has been characterized as a story of 'growth without development' (Easterly, 2001). Pakistan had a fractured economic inheritance as an outcome of partition and the post colonial development in a way saw a complete reconstruction of the economic and social infrastructure. Pakistan's growth pattern over the years therefore was contingent upon the level of external support that was extended to it under different regimes rather than there being a strong internal impetus to the growth process. In this scenario, the following became the key hall mark of Pakistan's economic evolution internally.

Poverty: Even though the country has experienced fairly rapid rate of economic growth in the recent times, the same has been unable to make a dent in the standard of living of majority of its people. The biggest indicator of this fact is the increasing incidence of poverty which has been well documented in the recent times. According to some recent estimates, the head count ration of poverty in Pakistan has increased from 33.8 percent in FY08 to 36.1 percent in FY09 (SBP, 2008). At the same time these figures have been contested and it is argued that the poverty levels may be much higher if one goes by some recent official estimates.

It is pointed out that around 62 million people may have fallen below poverty line, which amounts to nearly 42.8 percent of total population. Significantly, the share of 'severely food insecure population' is slated to have increased from 23 percent in FY06 to 28 percent in FY08 indicating the increasing vulnerability of the lowest rung. Increasing poverty carries within it seeds of distress as well unrest and has the potential for internal destabilization.

Demographic Factor: Pakistan has experienced the highest population growth rates in the recent years as compared to its other South Asian neighbours. It is considered to be the seventh largest country in the world in terms of its population size. What makes the situation a little more complex in case of Pakistan as compared to some of its other counterparts in the region is the fact that the demographic expansion has been the maximum in the age group of 15-45 years implying that the total numbers in terms of available productive work force is quite high.

Meanwhile, the available avenues for absorption of this teeming surplus have not grown commensurately, thereby making this huge chunk of population susceptible to tendencies of extremism and radicalism which may often offer easy material benefits. It has therefore been pointed out that there is a need for policies dealing with education, public health, and those that promote labour market flexibility and provide incentives for investment and savings. On the contrary, if appropriate policies are not formulated, the demographic dividend might in fact be a cost, leading to unemployment and an unbearable strain on education, health, and old (Nayab, 2006).

Growing Inter-provincial Disparities: Of the four provinces, Punjab scores high over the other provinces in terms of its growth and developmental indicators. Its economy has quadrupled since 1970s. It constitutes more than half of the total GDP of the country. In terms of its social indicators, literacy levels also are the highest in Punjab and the lowest in Baluchistan, which also corresponds to the participation of labour force in gainful employment, with Sindh and North Western Frontier Province in the middle. This inter-provincial disparity with dominance of Punjab is superimposed with the political domination the province has had with its contribution to the military-bureaucratic hierarchy. Such accentuating disparity was seen as one of the major impediments affecting political stability in Pakistan (Ishaque). In this context, even the decision of the civilian government for a Rs. 46.60 billion package for Baluchistan was seen as negating some of the basic premises of the problem which stems as much from alienation as from discrimination and deprivation.

Milibus in Pakistan: Given the overbearing role that military has had in shaping Pakistan's politics and security, it has always been known that it would need a deep economic penetration as well to sustain as well justify its role in the larger body politic. Over the recent years, Pakistan's military has created a comprehensive economic infrastructure for itself that not only symbolizes its presence in diverse sectors such as real estate, construction, education and manufacturing, but also seems to be giving it a *raison d'être.*

While its intervention in Pakistan's politics have been episodic and recurrent, military's role in economic enterprise in Pakistan has been a continuing but perhaps under researched story. Described as Milibus is one of the most authoritative writing in the area, it is argued, "The most serious consequence of military's involvement in economic affairs relates to their sense of judgment regarding political control of the state. The financial autonomy of the military reflected through their burgeoning economic empire, establishes the officer cadres interest in establishing political control over the state. Since political power nurtures financial benefits, the military establishment considers it beneficial to perpetuate it. In this respect, political and economic benefits are linked in a cyclic process: political power guarantees economic benefits, which in turn motivates the officer cadre to remain powerful and play an influential role in governance" (Agha, 2007).

What makes the situation more precarious perhaps is the fact that such deep economic penetration of military in economy comes at the cost of expenditure towards major social developmental concerns. While military expenditure as continued to be nearly one fourth of Pakistan's total expenditure, as a proportion of GDP it has been four times that of expenditure on health and nearly twice that on education in the 1990s (Chawla, 2001).

External Markers

In addition to the above internal markers in Pakistan's political economy, there are very deeply entrenched external markers which have had a bearing on regional security situation as well. As stated above, the two are not mutually exclusive, but have an overlapping and interrelated dimension. Some the critical external markers would include:

Pakistan and the Global War on Terror: It would not be an exaggeration to state that post-9/11 scenario brought back Pakistan's political economy form a state of near peripherality it was experiencing with the end of the cold war to a position of mainstream. Faced with no choice, Pakistan became a partner in the global war on terror in its immediate neighbourhood and was therefore bound to get affected by not only the successes of the campaign, but also its failures. A brief stocktaking of the impact of nearly ten years of the war on Pakistan puts some interesting figures.

On one hand Pakistan has been able to get a handsome financial and military help from the US and the Western donors to supplement its efforts. The US$ 1.5 billion annual aid pledged to the country under the PEACE project was seen as giving the much needed filip to the sagging economy. The IMF too was expected to boost its standby facility to Pakistan to US$ 12 billion from the existing US$ 7.6 billion in 2009. On the other hand it was felt that the aid actually coming in was miniscule in terms of the actual losses-both human and material that Pakistan has suffered in the war.

According to some estimates, the economy has suffered a loss of nearly US$ 7 billion annually in the aftermath of the war on terror with millions of people getting displaced internally and the total overall loss coming to a staggering US$ 68 billion, thus giving credence to the view that external help has been miniscule in relation to the losses suffered (Saleem, 2009).

More damaging perhaps has been the cumulative impact of the political economy of war that the war on terror has generated and the entrenched social and feudal structures that have sustained the same. As pointed out in one of the most recent formulations on the issue,

"The political economy of war and neo-liberalism, and a deeply embedded system of patronage, need to be understood if meaningful explanations are to be forged for why Pakistan is a hotbed of religious militancy, why a seemingly divided and militarised State remains powerful, and why substantive political transformation is unlikely in the short run" (Akhtar, 2011).

It is not surprising therefore that the recent pronouncements by the US administration to hold back US$ 800 million in military aid to Pakistan on account of its less than transparent dealings in the war on terror are likely to further complicate the matters rather than resolve them (*The Times of India*, 2011).

Global Recession and its Impact on the Economy: The onset of economic recession has affected Pakistan in the same manner as it has affected its other neighbours, but the intensity has perhaps been more on account of the political instability witnessed during the last few years. The crisis 'severely dented global aggregate demand, decimated liquidity in the international capital markets, and reduced investor confidence. The nation also witnessed a sharp decline in both portfolio as well as direct investments, and the access to international capital markets disappeared.

The decline in capital and financial account receipts (net foreign investment declined by 51.1 percent during FY09 on top of a 35.3 percent decline in the previous year), in the face of a widening current account deficit, led to a substantial depletion of the country's foreign exchange reserves, severe reduction in domestic liquidity, and a further impetus to domestic inflation' (SBP, 2008). There was related dimension of loss of jobs particularly in the exports sector, particularly the textile and garment sector (accounting for nearly 50 percent of the nation's total exports) which declined by 10.40 percent (Bashir, 2009).

Drug Economy: An area which is a rather under-researched one, drug production and trafficking is perhaps one of the most critical issues bearing on the bordering areas of Pakistan's North Western Frontier Province and the neighbouring Afghanistan. Pakistan has among the highest drug addicts in the world with some estimates showing that it equals 25 percent of the world's total. The total number was estimated to be three million addicts and their numbers are growing at a rate faster than the average population growth of Pakistan itself (Asad and Harris, 2003). With the increasing lawlessness in tribal areas and instability in Afghanistan, the region became the hub of opium cultivation and subsequent trafficking which brought easy money to fund the insurgent wars in the region.

India Factor: The last but certainly not the least factor in shaping Pakistan's overall posturing towards defence. Based on historical and ideological premises, the Indo-Pak rivalry in the South Asian region saw three full-fledged wars and decades of continuing acrimony over the Kashmir issue. The perceived threat from the 'hegemonic' India became pivotal in enabling Pakistan to erect a massive defence infrastructure which had to be ever ready to counter if not dominate the superior Indian conventional capability. The India factor also gave Pakistan the justification to excuse itself from global non proliferation and control measures such as NPT and CTBT.

The 1998 Nuclear test further brought the two nations in the global radar, where external mediation could be taken for granted lest the conventional clash takes the nuclear colour. This also enabled Pakistan to conveniently divert part of the material and financial support that was given to it to meet the Indian challenge. The India factor, not only enabled the military state to rationalize itself, it often became a rallying point for political opponents to be united.

The combined effect of these internal and external markers in the recent years has been to create a state apparatus where defence and military have become an end in themselves rather than means. The current civilian leadership in Pakistan which in any case is fraught with its internal contradictions remains a bystander when it comes to making a dent in the present defence architecture. The implications for South Asian Security therefore are obvious.

Implications for South Asia

Given the above markers, Pakistan remains one but significant factor in the larger context of South Asian security. The Indo-centric nature of South Asia will definitely make Indo-Pak relations central to any security considerations that need to be framed, yet one can think of some larger issues that will impact on South Asia based on the direction that Pakistan takes in the near future.

First, so long as the war on terror remains at the doorstep of South Asia, Pakistan will continue to figure in as a critical factor in shaping South Asian security. The argument derives on the inevitability of the external involvement so long as the war continues and since Pakistan is the only credible partner seen in the midst of the present turmoil (despite its numerous aberrations) centrality of the nation to any war effort shall remain.

Second, even if one presumes that there is going to be an external withdrawal from Afghanistan in the coming years, the situation seems highly unlikely to be resolved to ensure a stable frontier in the Northwest of the South Asian subcontinent. Pakistan shall continue to use its covert influences and leverages (as it did in the pre-9/11 days) to maintain its strategic depth in Afghanistan. This would certainly imply greater instability and turmoil in the neighbourhood.

Third is the question of internal political calculus in Pakistan. Given the recurrent pattern of military intervention in Pakistan and the continuing political conundrum, it may not be long before the military again starts preparing to reinforce its role as the mediating force and the balancer in Pakistan, even thought the underlying interests shall be dictated by the 'Military Inc'. The key question will be how the military will be able to resolve the growing restiveness and extremism in the tribal and provincial areas.

Fourthly, one is unlikely to expect a major breakthrough in so far as the Indo-Pak bilateral relations are concerned. The present state of bilaterals in any case appear to be conveniently placed on a cyclical path where a meaningful breakthrough is a foregone conclusion. It would not be an exaggeration to say that Indo-Pak talks shall remain a dialogue about dialogues rather than a meaningful breakthrough. Bigger players like US would intermediate only to the extent they would be motivated by their immediate concerns.

Finally, will it take the project of regionalism in South Asia any further? One is bound to be less optimistic as in the best of times, SAARC has not been able to achieve full scale participation of Pakistan as it has remained a reluctant partner in the grouping in the light of the 'reluctant leader' that India has been. SAARC is looked upon by Pakistan, more than any other nation as a mere forum to articulate its apprehensions regarding the bigger neighbour and less for a constructive regional engagement.

Conclusion

It becomes apparent therefore that political economy of contemporary Pakistan, shaped by its internal and external markers is likely to make it a critical factor in South Asia's stability. But Pakistan's security calculus in the unfolding scenario in South Asia will continue to be shaped by the perceptions and inhibitions of the past rather than any meaningful breakthrough for future. South Asia therefore must define its own security priorities from a regional perspective lest it allows the external to predetermine the agenda. The sooner it does it the better.

References

Agha, Ayesha Siddaqa (2007): *Military Inc. Inside Pakistan's Military Economy*, Oxford: Oxford University Press, p. 257.

Akhtar, Aasim Sajjad (2011): Peering into Pakistan, *Economic and Political Weekly*, July, 9, LVI (28): 31.

Asad, Amir Zada and Robert Harris (2003): *The Politics and Economics of Drug Production on the Pakistan-Afghanistan Border*, Hampshire: Ashgate, p. 124.

Bashir, Rizwana (2009): "Fallout of Political Instability: Declaration in Exports", *Pakistan and Gulf Economist,* April 13-19, p. 15.

Burki, Shahid Javed (2008): *Changing Perceptions, Altered Reality: Pakistan's Economy Under Musharraf, 1999-2006,* Karachi: Oxford University Press, p. 348.

Chawla, Shalini (2001): "Pakistan's Military Spending: Socio Economic Dimensions", *Strategic Analysis,* 25 (5): 714. While military expenditure as a percentage of GDP was 4.8 percent in 1998-99, it was 0.7 percent for health and 2.2 for education.

Easterly, William (2001): 'Political economy of growth without development: A case study of Pakistan', *Paper for the Analytical Narratives of Growth Project,* Kennedy School of Government, Harvard University, June.

Haider, Syed Fazal, "Stabilizing the restive Balochistan for Economic Development", *Pakistan and Gulf Economist,* April 13-19, p. 13.

Hussain, Akmal (1998): *Strategic Issues in Pakistan's Economic Policy,* Lahore: Progressive Publishers, p. 364.

Hussain, Ishrat (2009): "The Role of Politics in Pakistan's Economy", *Journal of International Affairs*, 63 (1).

Ishaque, Fozia "Interrelaint Realms-Politics and Economy", *Pakistan and Gulf Economist,* April 13-19, p. 12.

Mohammad, Waseem (1994): "Politics and the State in Pakistan", Islamabad: National Institute of Historical and Cultural Research.

Nayab, Durr-e- (2006): Demographic Dividend or Democratic Threat in Pakistan, *MPRA Paper No.2227,* page 5, online at http://mpra.ub.uni-muenchen.de/2227/

Saleem, Kanwal (2009): "Political Instability, Poor Law and Order Hitting Hard to the Economy", *Pakistan and Gulf Economist,* April 13-19.

State Bank of Pakistan (2008): *Annual Report,* Chapter 1 Overview, p. 1.

The Times of India (2011): 'US Stalls Military Aid to Pakistan', 11th June, 2011.

12

Whither India's Bomb? Motivational Models Revisited

Rajesh Basrur

In 2008, there was considerable interest in the state and possible future of India's nuclear weapons program and strategy. There were two reasons for this. First, the year marked the tenth anniversary of the tests conducted by India in May 1998 (and by Pakistan shortly thereafter) and there was considerable debate over the lessons of the decade that had elapsed (Ganguly, 2008: 45-70; Kapur, 2008: 71-94; Krepon, 2008). Second, a more protracted and furious debate among the closed circle of nuclear sellers over the merits of allowing the transfer of nuclear fuel to India came to a peak, with opponents arguing that access to international supplies would enable India to divert its own stocks of fissile material for the production of more nuclear warheads (Fitzpatrick, 2008: 76-85; Paul, 2007: 845-861; Weiss, 2007).

The first debate was in a way misleading, since it tended to add to the commonplace tendency to conflate India's nuclear capability with its declared intention. Indeed, India had demonstrated its capability much earlier in 1974, but had refrained from building an arsenal for well over a decade. The actual construction of the Indian bomb sometime in 1989 came nearly a decade before the 1998 tests, when India officially declared itself to be a nuclear-armed state. The long stretch of time encompassing the capability, the making and the declaration encompasses a complex story that helps answer the question central to the second debate. Where is India's nuclear weapons programme going two decades after the making of the bomb?

The question merits attention because, for all its current limitations as a poor and still developing country with limited capacity to project power, India's rise as a significant power with a strategic reach extending beyond the subcontinent is now a widely acknowledged reality (Baru, 2006; Pant, 2007). What role India will give to its nuclear programme will have an important impact upon the security politics of an increasingly Asia-centric world. Will an ambitious India expand its nuclear capabilities significantly and thereby generate new tensions with another expanding military (and nuclear) power, China? Or will it retain its current low-key nuclear posture, refrain from actively deploying its weapons, and perhaps even play a leading role in the current drive to engineer a post-nuclear era? The signals it has been sending out are conflicting. First, India has ridden out severe crises with Pakistan in 1999, 2001-02 and 2008-09 without displaying its nuclear wherewithal.

On the other hand, its research and development program envisages the production of a wide array of delivery vehicles designed to achieve a triad of air, land and sea nuclear forces. Second, and more fundamentally, while the official doctrine of minimum deterrence is backed by a recessed or non-deployed posture and a commitment to disarmament, the language of Indian nuclear doctrine resting on the American lexicon of deterrence that revolves around "assured second strike capability" is conducive to a wide expansion of its arsenal, especially under pressure of competition and strategic tension. Third, within its own neighbourhood, South Asia, India's role has been so restrained as to bring it the sobriquet of "reluctant hegemon" (Mitra, 2003: 365-380). Yet in its quest for permanent membership of the UN Security Council, its expanding strategic role in Asia, and its rapid military modernization program, it has displayed all the expected preferences for the position, status and regalia of a would-be great power (Giridharadas, 2008).

This paper will attempt to dissolve the ambiguity and

clarify where India's nuclear weapons programme is headed. But to better understand where India is going, it is important to have a sense of where it is coming from. The three key points in its history noted above—the 1974 test, the construction of the bomb in 1989, and the 1998 tests are particularly important in attempting to gauge the motives driving the programme and where it stands today. The literature on motivation/causation why do states seek the bomb?-provides the beginnings of an analytical framework. The three models identified by Scott Sagan—the security, the domestic politics, and the norms models—provide a useful starting point (Sagan, 1996/97: 54-86). In this paper, I apply them to a historical process rather than to a single decision; refine them internally to bring out the effects of their diverse components; and taken into account their cross-connections to produce a more nuanced and complex explanation of the forces driving India's nuclear weapons program over time. Taken together, the trajectories produced by each model bring out a clearer picture of what lies ahead.

At each point, as will be seen, the motives were somewhat different and the driving forces and their relative weights varied, while the action taken was consistently restrained. This shows, I argue, an overall trajectory of limited expansion and strategic restraint that leads to the expectation of moderation in times to come. India is likely at most to continue along the path of strategic drift driven by uncertainty about the requirements of the minimum deterrence doctrine it espouses. This augurs the leisurely development of a range of capabilities, but continued adherence to a non-deployed posture. On the other hand, India is likely to retain its interest in—and may even play a significant role in—the global drive to reduce and marginalize nuclear weapons with a view to eventual elimination. In short, it is unlikely to treat nuclear weapons as a usable instrument for attaining great power status.

In the next section, I develop an analytical framework that

begins with Sagan's models of motivation, refines them, and shows how they may be integrated into a single explanatory framework. In the subsequent three sections, I apply each of the refined models in succession to assess Indian motivations through the time span of India's nuclear weapons history. In the final section, I pull the separate threads together to weave an assessment of India's future course as a nuclear power, bring out the implications of the study for parallel assessments of the motivations of other nuclear-armed states and, more broadly, suggest ways in which this analysis can help us theorize about motivations.

Analytical Framework

A large number of explanations for India's motives in going nuclear have been offered, with the largest number, though not all, focusing on the 1998 tests (Cohen, 2000; Hagerty, 1999: 19-29; Jha, 1998: 80-96). These can be classified as falling under one of three categories or models identified by Scott Sagan in his widely cited work on nuclear motivations: the security model, the domestic politics model, and the norms model (Sagan, 1996/97). The first corresponds to the systemic level of analysis and the other two to the state level, with the individual/small-group level of decision making subsumed by it. In this section, I briefly discuss the major writings on each level, and provide an initial glimpse of the nuanced picture that will follow in the next three sections.

The security model, which corresponds to the realist perspective on inter-state relations, rests on the view that, in an anarchic international system, power—and especially military power—is the key determinant of state survival. [1] States that feel threatened by others are inclined to seek appropriate military instruments to try and obtain insecurity. It follows that an India fearful of the nuclear threat posed by its adversaries, would pursue nuclear capability.

This does not, however, necessarily tell the full story. Threats that drive a state to nuclearise may be of different kinds. It is useful to distinguish between immediate threats and prospective ones that may or may not become serious—a distinction parallel to that between immediate and general deterrence (Morgan, 2003: 80-86). I do not, after all, lock my door only if a burglar lurks outside; the possibility of being burgled at some unknown point in time is sufficient reason for doing so regularly. To carry the metaphor forward, I will view any effort to prevent me from choosing to lock the door when I consider it appropriate a security threat as well. In other words, a state which finds its choice–of acquiring nuclear weapons or not–blocked will regard the constraint imposed on it as an infringement on its strategic autonomy. These distinctions are important in understanding India's concerns and responses over time, I will show.

The domestic politics model, typically, is concerned with the ways in which the configuration of domestic political power shapes the making of policy decisions. A specific decision may be the result of a bargain between competing groups or of a condominium between or among powerful interests. Those likely to be the producers of weapons or their custodians and potential utilisers are bound to have a vested interest in pressing for nuclearisation. The nature of the political system may have a bearing on the outcome.

For instance, in a system where decision-making is concentrated, the personality of individual leaders is likely to influence a decision. [2] Contrary to common misconception, this applies to democratic states as well. In Nehru's India, for instance, the prime minister dominated decision making in foreign and security policy. It has also been argued that a political system that is more open to the world—particularly with respect to integration with the global economy—is more likely to be aware of the negative repercussions of obtaining nuclear weapons and hence to abjure them (Solingen, 2007).

How far these aspects of domestic politics have influenced the making of Indian policy and strategy will be examined below. For the moment, it is important to bear in mind that political bargaining is not necessarily to be understood as an on-going tussle over what is to be done. There may also be a stable bargain over time that enables the weaker of the two sides to exercise considerable control over the domain granted to it as its agreed sphere of influence. This, as I will show, is the reason why the armed forces in India have a significant degree of influence over nuclear weapons strategy, but not posture.

The norms model presents nuclear weapons as symbols—of power, modernity, and prestige—that shape a nation's identity. Their symbolic meaning may change, as Sagan notes: in the 1960s, nuclear testing was considered "prestigious and legitimate," but today it is "illegitimate and irresponsible" (Sagan, 1996:76). More broadly, a norm-driven pattern of thinking and behaviour or strategic culture–developed over time may mould a state's nuclear weapons programme in significant ways. [3] Finally, the meanings we attach to words that make up our vocabulary may play a vital role in determining how we act. A large part of the discourse on nuclear weapons, especially on the central term "deterrence," is symbolic—shaped by expectations and beliefs rather than by actual experience (Jervis, 1989: 174-225). As will be seen below, the meanings attached to the weapons may be different for the different groups connected with them, thus allowing for the coexistence of contrary beliefs and attitudes about the weapons.

The three models are not exclusive, for there are clear linkages across models. The security model is influenced by the domestic politics model in at least two ways. First, if there are conflicting interpretations of a security threat requiring a possible nuclearisation response, the determination of the policy response depends on whose influence prevails, i.e. on the political equation of the time. Second, the nature of the response—the strategy adopted and the mode of its implementation—depends again on who influences the response, and that may not necessarily be the same group as the first. The distinction, I will show, is important. The decision to cross the nuclear Rubicon is one thing; what is to be done on the other side is quite another. Both of these outcomes also depend on the norms espoused by the concerned groups.

Given the use of three distinct models, how is an integrated explanation to be developed? In this paper, I draw from a comparable exercise by James Mahoney and Richard Snyder, who attempt to integrate multiple levels of analysis within the larger context of the agent-structure debate. [4] The "funnel strategy" identified by them involves a unidirectional movement from the broadest level (the systemic) to the narrowest (leadership). The key to integration is the identification of vectors at each level. The outcome is explained by "summing forces and directions of variables" (Mahoney and Snyder, 1999: 12). In the present exercise, because the models do not strictly correspond to distinct levels of analysis, there is no need to go down the "funnel." It is sufficient to identify the trajectory or vector for each model and observe the summing of the three to try and answer the question posed at the outset: where is India's nuclear weapons programme going?

The Security Model

The security model applies to the entire period of India's nuclear history, but in different ways at different points of time. In the years after Independence (1947), Jawaharlal Nehru, India's first prime minister, was the predominant figure in shaping security policy. Though deeply troubled by the genocidal possibilities inherent in the bomb, he kept his options open as he was conscious of the "built-in advantage" of deterrence the technology offered should the need arise (Kapur, 1976:193-194). Soon after it had defeated India in the Himalayan war of 1962, China tested its first nuclear weapon in 1964. This prompted Prime Minister Lal Bahadur Shastri to order research and development toward a nuclear test, though he limited its ambit.

Thus, the motivation shifted, though not sharply, from a very broad general deterrence approach to something a little closer to immediate deterrence vis-à-vis a specific adversary within a narrower time frame. In 1974, Indira Gandhi carried out a single test from a different security perspective. Sagan has argued that the test decision was taken in "haste," and that it appears to have been motivated by domestic considerations as Mrs. Gandhi's political fortunes were ebbing (Sagan, 2007: 68). But in fact the decision had been taken much earlier somewhere between December 1971 and September 1972 (Perkovich, 1999: 172).

At the time, India had fought a successful war with Pakistan but only after a period of intense uncertainty when it was virtually isolated by what was viewed as an emerging US-China-Pakistan nexus in response to which Mrs. Gandhi had had perforce to enter into a quasi-alliance with the Soviet Union in August 1971. The 1974 test was a symbolic declaration of Indian autonomy. Though Mrs. Gandhi was not interested in going nuclear, which explains why she did not build the bomb, she used the test to assert India's capability and strategic autonomy. [5]

During the 1980s, as evidence of Pakistan's quest for nuclear weapons began gradually to appear, the focus shifted again to the need for immediate deterrence owing to the on-going tensions between the two countries. In the Indian perception, there were two secondary contributors to the threat. There was evidence that China, pursuing a strategy of containing India by using Pakistan as a surrogate, had supplied technology and possibly nuclear materials and warhead designs for the Pakistani bomb (CNPS, 1999). The United States, meanwhile, had ignored growing evidence of Pakistan's quest as it needed Islamabad's close cooperation in building up the armed resistance of the mujahideen to the Soviet occupation of Afghanistan. Faced by a direct military threat, a reluctant Rajiv Gandhi ordered the construction of the bomb in 1989, though it was to remain in the basement for nearly a decade (Subrahmanyam, 1998: 44).

The 1998 tests carried out by the government of Prime Minister Atal Behari Vajpayee have been variously attributed to domestic politics and the "Hindu nationalism" of Vajpayee's Bharatiya Janata Party (BJP), but the critics fail to acknowledge that Vajpayee had merely carried out a plan premiers from other parties had prepared for but had hesitated to implement. [6] For instance, Prime Minister Narasimha Rao had ordered a test in 1995, but refrained from carrying it out under US pressure.

The real concern in this period was security in the sense described above as potential loss of decision making power: the fear that India was about to lose its strategic autonomy. Flushed with its Cold War triumph, the United States had begun to squeeze India and to try and "roll back" its nuclear weapons option. Indian leaders were particularly incensed by the entry into force clause of the Comprehensive Test Ban Treaty (CTBT), which envisaged putting pressure on non-signatories to toe the non-proliferation line. The 1998 tests thus represented a breakout against the attempt of the US-led non-proliferation regime to close India's nuclear options permanently (Saksena, 2006: 209-229).

However, the bomb was not of much use against an implacable adversary determined to exploit its strategic potential in new ways. Pakistan, having obtained its own nuclear capability sometime during the 1980s, had begun to exert pressure on India through sub-conventional means. Confident that its own bomb deterred an Indian conventional response, it stepped up its support for terrorist groups attempting to sunder that portion of the disputed territory of Kashmir held by India (Kapur, 2007). The 1990s saw a wave of terrorist violence in India that brought home to its leadership the problem of the "stability/instability paradox," in which mutual nuclear deterrence encourages strategic instability at a lower level of conflict. [7]

The problem became worse over time as a series of crises engulfed India-Pakistan relations (Chari, Cheema and Cohen, 2007; Ganguly and Hagerty, 2005; Kapur, 2007). In 1999, Pakistani troops in mufti occupied large tracts of land on the Indian side of the Line of Control (LoC) that divides Kashmir and a short but intense period of fighting resulted in high casualties on both sides. In 2001-02, after terrorists attacked the Indian parliament, India threatened to punish Pakistan and mobilized massively for war, though it eventually backed down. In November 2008, Pakistan-based terrorists attacked the Indian city of Mumbai and killed about 170 civilians, which produced yet another India-Pakistan crisis (Hoyt, Hussain and Mandal, 2009; Rabasa et al., 2009).

The hard lesson for India has been that nuclear weapons provide security against nuclear attack, but provide adversaries with incentives for aggressive action at lower levels of violence. In effect, a new kind of security dilemma comes into play. In the traditional conception of the security dilemma, a state is caught between not arming and arming because the former leaves it open to attacks by adversaries, whereas the latter produces a like response that leads to arms racing and spiralling tension (Jervis, 1978:1167-1214; Booth and Wheeler, 2008; Glaser, 1997:171-201). In the case under discussion, India has been caught between the insecurity of not arming and the insecurity produced by the stability/instability paradox.

A related lesson is that nuclear deterrence is easily obtained. Both India and Pakistan have a handful of relatively unsophisticated weapons that have been kept unassembled and have never been displayed during a crisis, but have yet deterred them. Indeed, even in the relatively minor crisis of 1990, while India had no clear information on Pakistan's nuclear capability, interviews show that Indian leaders were "alarmed" because a war "might be a nuclear war" (Chari, Cheema and Cohen, 2003:139-140). In subsequent crises, which occurred after Pakistan's 1998 tests, though its capabilities were never displayed, that awareness grew (Chari et al, 2003: 139-140). It follows that programmes for building large arsenals would only involve pursuing superfluous capabilities.

To recapitulate, the Indian historical experience has been diverse. Rather than the single-minded pursuit of nuclear capability against threats posed by an individual state, it has involved securing the Indian state's strategic autonomy and making minimal and (in purely nuclear terms) adequate preparations to counter direct threats to survival. In the process, while nuclear deterrence has been obtained at a low level, even at that level it has been evident that nuclear weapons have not brought unadulterated security with them. From both points of view, the vector points toward restraint, for there is no incentive to seek greater capability.

Domestic Politics Model

The domestic politics model deconstructs national interests and looks at who defines the national interest as well as at the sectional interests of individuals and groups able to influence policy (Gourevitch, 2002). This requires an understanding of the way a political system functions, especially how competitive politics is conducted and the kinds of political bargains that shape policy outcomes. As noted earlier, in the present context, the role of individuals and their personalities is also closely related to the configuration of the political system. In this section, I analyze the impact of the constitutional design of the Indian political system and the way it functions, the ways in which individual leaders shape policy making, and the sectional interests and relative weights of key groups connected to the nuclear weapons programme.

With regard to the nature of the political system, Etel Solingen makes the interesting argument that outward-looking elites tend not to favour the acquisition of nuclear weapons because of the potential costs (such as economic isolation) whereas autarky-oriented elites, less concerned about such costs, are disposed to nuclearise (Solingen, 2007). The pattern does not fit well in the Indian context. Indira Gandhi, who may be classified as inward-looking, did carry out a test in 1974, but she refrained from building the bomb thereafter. In contrast, Rajiv Gandhi, whose prime ministership is associated with a more outward orientation, ordered the construction of the bomb.

In the early to mid-1990s, Narasimha Rao's Congress, which had shifted dramatically to economic liberalization, did avoid testing, which appears to confirm Solingen's thesis, but it retained the bomb. While the BJP had before coming to power aired the slogan of "swadeshi" (best transliterated as "made at home"), which carried the flavour of autarkic nationalism, the 1998 tests came at a juncture when the government led by it *accelerated* the process of deregulation and liberalization of the economy. Clearly, the Solingen thesis does not apply.

The constitutional organization of Indian politics has had a bearing on the making of policy generally and nuclear policy in particular. Unlike the US system, where the division of powers clearly separates the executive from the legislature and necessitates a bargaining process over policy, India has a parliamentary system in which there is no such clear division: the executive headed by the prime minister enjoys its power by virtue of its majority in (and hence control over) the legislature. Though policies may be debated extensively, there is rarely a head-on clash between the two branches. The scope for lobbying by interest groups and for bureaucratic politics is much narrower whereas the role of the individual leader—mainly, the prime minister–in determining the direction of policy is relatively strong.

In this context, the personality and preferences of the prime minister are important in shaping policy. Notably, successive prime ministers have always held direct control over the Departments of Atomic Energy and Space. The broad thrust of Indian policy vis-à-vis nuclear weapons was given by

Nehru, who–as noted above–was reluctant to develop nuclear weapons though also unwilling to close the option. Despite testing in 1974, Mrs. Gandhi too refrained from exercising the option, though she did widen the option when she launched the missile programme in 1983.

In a widely appreciated work, Jacques Hymans has tried to show that certain personality types can be associated with the decision to nuclearise (Hymans, 2006). Specifically, he argues that most Indian leaders were "sportsmanlike nationalists" with a preference for autonomy but not for acquiring the bomb, whereas Vajpayee was an "oppositional nationalist" who was predisposed to go for the bomb.

However, this typology does not quite work in the light of evidence. First, as Hymans acknowledges, Rajiv Gandhi ordered the fabrication of the bomb even though he was a sportsmanlike nationalist. Hymans tries to qualify this by arguing that "what is really striking is not that Rajiv Gandhi moved toward nuclear weapons, but that he did not decide to go all the way" (Hymans, 2006: 191). But the qualification shifts the linkage from the decision to go nuclear to a decision about how far to go.

More problematic is the classification of Vajpayee's personality. Hymans cites a poem written by the young Vajpayee to illustrate the fire of spiritual rage he harboured, calling this a "chilling premonition of the tests he was to order" (Hymans, 2006: 202). This characterization of Vajpayee's personality is starkly contradicted by other facts. First, as a poet, he had also expressed his deep moral anguish about nuclear weapons, calling them "monstrous" and wondering whether history would ever forgive their inventors. [8] Second, as minister for External Affairs in Prime Minister Morarji Desai's cabinet after the 1977 elections, Vajpayee had voted *against* the nuclear option on precisely the grounds Hymans associates with sportsmanlike nationalists: the fear of generating fresh insecurity (Subrahmanyam, 2004). Third, though he appeared to be a hawk, Vajpayee repeatedly offered an olive branch to Pakistan, notably when he travelled to Lahore in 1999, met General Pervez Musharraf at the Agra summit (2001) and set in motion a peace process in 2003. While the role of the individual is hard to pin down through typologies, it still remains true that prime ministerial preference, especially in the early years after independence, established a pattern of restrained interest in nuclear weapons. I will elaborate on this pattern in the following section.

The role of the scientific community within the government—the technocrats—has been the subject of considerable scholarly research (Abraham, 1998; Abraham, 1997: 2136-2146). Given the environment of secrecy, their lack of public accountability and their direct access to the prime minister, India's nuclear scientists and engineers were able to exercise considerable influence in pursuing the interests of their bureaucratic empire.

However, it is quite another thing to show that the technocrats *drove* nuclear policy. On the contrary, there is sufficient evidence that the opposite was the case: that it was the political leadership, even when the prime minister was weak, that played the decisive role. Homi Bhabha, who established the Indian nuclear programme and was a close friend of Nehru, could not persuade the latter to think seriously of pursuing a weapons program (Chengappa, 2000: 88-89). Indira Gandhi changed her decision to test in 1982 without consulting the so-called "strategic enclave" (Chengappa, 2000: 260-261). In 1995, Narasimha Rao rejected strong technocratic pressures to test (Chengappa, 2000: 393). In 1996, the relatively weak Prime Minister, H.D. Deve Gowda, overruled the atomic energy establishment's insistent demands for testing (Chengappa, 2000: 397).

Ironically, in more recent times, the political influence of the nuclear scientists has increased, but its involvement in the politics and debates surrounding nuclear weapons has diminished. Both aspects were visible during the prolonged and intense debate over the India-US nuclear agreement. The scientists' influence was enhanced by the weakness of the Congress-led coalition (comprised of nearly two dozen parties) and the internal divisions between the Congress and its allies, notably the Communist Party of India-Marxist (CPM). [9]

Fearful of incursions onto their turf as a result of US-imposed conditions, the nuclear scientists forged an alliance with the CPM and successfully staved off external control over their empire. [10] Insofar as the nuclear scientists have a parochial interest, that lies in the realm of civilian energy, which is gearing up for a major programme of expansion that could amount to a quadrupling of nuclear energy production by 2020. [11]

Notably, the publicity given to Indian achievements with respect to nuclear weapons has been generated largely by the engineers of the Defence Research and Development Organization (DRDO), a sprawling public sector conglomerate that produces missiles for the nuclear forces and a wide range of conventional military hardware.

A less noticed influence has been the military. On the face of it, this seems unlikely, for the Indian armed forces have long been under the tight control of their civilian masters. Yet the influence of the military is considerable because of the nature of India's civil-military bargain, which mandates strong civilian control, but leaves the military largely autonomous in determining the requirements of defence—in this case, of deterrence. [12] Whiie final decisions are made by the civilian authority, it is the military which decides what the operational requirement of deterrence is. Its approach is generally expansive in tenor, in good part due to its doctrinal understanding of the working of deterrence

If the military determines that a large, sophisticated and diversified force is essential for national security, it has a fair chance of influencing acquisition policy. Thus there appears to be an unobtrusive "alliance" between the armed forces and the delivery vehicle producers which, despite their official subservience to the civilian leadership, is pushing the expansion of the Indian nuclear weapons programme.

In part, the influence of the military has occurred by default: the civilian arm of the state—both its political and bureaucratic branches—has not troubled to understand the language and workings of nuclear deterrence. While declaring its adherence to minimum deterrence, it has tended to permit a drift toward a more expansive understanding of the concept (Basrur, 2008). As a result, Michael Kraig observes, there are signs that civilian leaders and analysts are "buying into" the military's assumptions and reasoning (Kraig, 2003: 30).

However, this criticism should not be taken too far. First, the government is clearly aware of the need to avoid cutting the military too much slack, for which reason it has steadfastly prevaricated on the issue of appointing a Chief of Defence Staff (CDS) as suggested by the recommendations of a Group of Ministers report in 2001. [13] Second, it has consistently adhered to its preference for non-deployment of its forces in what is indubitably a highly minimalist posture. And third, with the primacy given to economic development, the state has not shown much interest in building a large arsenal. That said, the tension between the restraint exercised by the civilian leadership and the expansion-oriented views of the armed forces remains.

Overall, notwithstanding these conflicting tendencies, the vector points toward limited expansion. The political leadership remains in control, retains a minimalist conception of the requirements of deterrence, abides by a recessed or non-deployed posture, and is aware of the risks of military domination of policymaking. The military is tightly controlled by the civilian authority. The nuclear scientists have never been unduly influential, but in any case have less interest in military matters than they did in the past. The missile makers have no significant political clout and are under review for an anticipated revamp of the organization owing to its many delays and cost overruns (Pubby, 2009).

Norms/Symbolic Model

This model highlights the symbolic aspects of nuclear weapons, though—as noted earlier—this may be favourable to nuclear weapons in some ways and antithetical to them in others. In this section, I analyze the normative and symbolic aspects of nuclear weapons in three respects. First, I evaluate the view that nuclear weapons are viewed as desirable because they are symbols of prestige. Second, I assess the view that nuclear weapons are in fact the opposite–that they are regarded as undesirable because they are objects of fear and anxiety. And third, I examine the power and effects of nuclear discourse as a symbolic discourse.

It is arguable that nuclear technology–as a cutting edge technology of the post-Second World War era—has a distinctive non-military appeal. As Itty Abraham holds, it was in the early days after independence a symbol of modernity for a country that had just emerged from a long period of colonial rule. Accordingly, it exercised a strong appeal for a leadership determined to "transform the landscape of modern India through the power of modern technology…" (Abraham, 1998: 20). Did this symbolism extend to nuclear weapons? Some analysts argue that it did, but focus mainly on the cultural nationalism of the BJP (Frey, 2002).

However, there are three objections to this. First, there is no necessary contradiction between gaining prestige and pursing security. On the contrary, it should be self-evident that success in obtaining security naturally raises prestige. For a realist, prestige is a vital component of security (Gilpin, 1981). Indira Gandhi's symbolic test was designed precisely to achieve this end. Second, as noted earlier, the cultural nationalism of BJP leaders is not an adequate explanation for the 1998 tests since Vajpayee had opposed testing in the late 1970s. And third, the prolonged reluctance of Indian leaders to permit the construction of the bomb is attributable to an opposite norm: a norm *against* nuclear weapons as symbols of prestige or sources of assured security.

The freedom movement and the first decades after independence were dominated by leaders like M.K. Gandhi and Nehru who were reluctant to accept the security value of nuclear weapons, but instead viewed them as sources of insecurity and moral abhorrence (Anand, 1998; Basrur, 2001: 181-198). The starkest examples have been cited already: Mrs. Gandhi, the quintessential realist who calculatedly broke up Pakistan in 1971, did not pursue the bomb; and Vajpayee, the "Hindu nationalist" whose party had always called for nuclearisation, blocked testing two decades before he actually ordered it. In between, Rajiv Gandhi, under whose orders the first Indian nuclear weapons were constructed, was a strong advocate of universal nuclear disarmament, declaring publicly in 1988 that the notion that nuclear weapons keep the peace is "false" (Chengappa, 2000: 330).

The essence of the Indian approach to nuclear weapons is brought out in the one official document on strategy released in January 2003—a laconic press release that announced among its chief features a policy of no first use (NFU) and commitment to disarmament. [14] After the 1998 tests, the BJP-led government and its Congress-led successor adhered to a minimalist non-deployed posture despite a series of crises with Pakistan in 1999, 2001-02 and 2008-09. The BJP's Minister for External Affairs, Jaswant Singh, also declared that "parity is not essential for deterrence," that "we do not see nuclear weapons as weapons of war fighting," and that a "triad" of delivery vehicles (air, land- and sea-launched) is not a "pre-requisite for credibility" (*The Hindu*, 1999). All of this points to a normative preference for nuclear minimalism and a bias against arms racing and open-ended nuclear expansion.

Yet there is a built-in tension arising from conceptual ambiguity about the nature of deterrence. Jaswant Singh, in the interview cited above, also said that the "minimum" in minimum deterrence cannot be a "fixed physical quantification," but is a "dynamic concept" that is "firmly rooted in the strategic environment, technological imperatives and national security needs…" (*The Hindu*, 1999). This clearly implied, in direct contradiction of his rejection of parity as a prerequisite for assured deterrence, that security is relative to the capabilities of others, and that the requirements of deterrence have to be open-ended.

This reflects a commonplace tendency among Indian strategic analysts to draw from the Western (mainly American) conception of deterrence based on "assured second-strike capability," which is rooted in conventional balance of power thinking—a realist proclivity that has no basis in the way that nuclear deterrence actually works, since in practice large(r) forces have never given any nuclear power a strategic advantage. [15] In practice, small forces have always deterred large ones, as is most clearly evident in the Cuban Missile Crisis as well as in other confrontations between nuclear-armed states. [16]

Despite the discrepancy between the discourse on how deterrence is said to work and the reality of how it actually works, the former holds sway among those concerned with the operational aspects of nuclear weapons. Thus, the shibboleth of "assured second-strike capability," which is a "dynamic concept" and logically leads to an open-ended search for security and to arms racing under the security dilemma, is ubiquitous in the writings of Indian nuclear strategists. [17] Why this discrepancy is allowed to persist can only briefly be explained here. In essence, nuclear politics is to a large extent symbolic politics. Because the world lacks experience with real nuclear conflict, the notion of what is real grows out of beliefs (Jervis, 1989: 180-185).

Though, post-Nagasaki, nobody has knowledge of (or even wants) the use of nuclear weapons, operational requirements necessitate the development of a discourse about how to use them and equally, about how to prevent their being used. This nuclear "thought style" involves a high degree of simplification and treating nuclear weapons as if they were conventional weapons. This is reflected in the experts' determination of the numbers and types of weapons considered to be adequate, the importance they give to calculations of relative damage, and their estimation of the capabilities required (O'Neill, 1999). The mental processes that permit the gap between beliefs and the reality of deterrence to be sustained involve a kind of sleight of mind known as "cognitive heuristics" or mental shortcuts commonly used to avoid being bogged down by complexity (Kanwisher, 1989: 652-675). The absence of real knowledge is substituted by "fantasy knowledge," which, moreover, hinders the absorption of reality-congruent knowledge (Elias, 1991: 57-58, 72).

A close adherence to American doctrinal discourse is visible, for instance, in the "Draft Nuclear Doctrine" produced in 1999 by India's National Security Advisory Board (NSAB), which calls for a triad of air, land and sea-based delivery systems whose "survivability will be enhanced by a combination of multiple redundant systems, mobility, dispersion and deception." [18]

Similarly, the unacknowledged absorption of the premises of the American doctrine of flexible response (the ability to fight at all levels) is manifest in the Indian Army's "Cold Start" doctrine, which envisages rapid mobilization for limited war (Ladwig III, 2007/2008: 158-190). The mantra of "survivable forces" is regularly recited by strategic experts, weapons manufacturers, and the media to justify an expanding research and development agenda for nuclear delivery systems. The power of this symbolic discourse lies in the autonomous space accorded to the military by the civil-military bargain referred to above. Since the military has the power to determine its operational requirements, its expectations are geared to expanding capabilities. The extended requirements are grist to the weapons producers' mill and hence the development of an array of missiles of various ranges as well as the quest for the sea leg of a planned triad.

We have seen here the tension between the restraint arising from the beliefs of the political leadership and the compulsion to expand arising from the operational pressures of an opposite belief system embedded in a powerful symbolic discourse. So far, the contradiction has been sustained without either side giving in, but it is certainly possible that policy makers will come under pressure to lean toward the operational perspective. How will they respond then?

In all probability, the outcome will be determined by the normative preferences of the political leadership, which does not have an operational perspective on nuclear weapons and which has full control over the weapons producers and managers.

Two points are notable. First, the political leadership has invariably retained tight control over the weapons operators and exercised great caution in times of crisis; and, second, during normal times, the civilians have shown a strong preference for a basically non-offensive nuclear posture that requires nuclear weapons to be stored in de-mated condition. In sum, the vector in this section points to restraint.

Conclusion

The preceding discussion of the three models commonly used to explain nuclear motivations shows that, in all three cases, the vectors point in the same direction, i.e. toward nuclear restraint. Thus there is a high probability that the Indian nuclear weapons program will continue to develop a wide range of operational capabilities at a pace dictated by its operational stakeholders, but restricted by the norms of the political leadership. Though capabilities may grow (slowly, owing to limited resource allocation), India will retain its non-provocative non-deployed posture. Nuclear crisis experience has clearly strengthened the minimalist understanding of nuclear deterrence. Indian troops did fight Pakistani forces in 1999, but only on the Indian side of the LoC.

Neither side officially violated that red line. In 2001-02, notwithstanding its assertion that limited war was feasible, the Indian government refrained from any military action and even punished senior officers whose actions risked involuntary escalation. In the crisis that followed the Mumbai terrorist attacks in November 2008, there was no serious effort to consider armed combat. In February 2009, Minister External Affairs Pranab Mukherjee publicly said that, though the attacks had originated in Pakistan, military response was not an option. [19] Subsequently, during the campaign for the general elections of April-May 2009, he even claimed that his diplomacy had worked better against Pakistan than the Vajpayee government's focus on the bullet. [20] Reading between the lines, it is clear that the lesson has been learned that military means cannot be relied on to solve India's security problems.

Nevertheless, it is useful to consider under what conditions India might opt for a rapid expansion of its arsenal and shift to a deployed force. The security model would look for external threats as a source of policy change. Externally, India might respond to a powerful stimulus from an adversary in the form of an aggressive nuclear build-up. Of the two conceivable adversaries that it will face in the foreseeable future, though it has shown signs of wanting to forge ahead of India quantitatively, Pakistan is limited by its much smaller economy. [21]

China is another question altogether and will be watched carefully by Indian strategists and policy makers, particularly in the context of its relatively high defence spending and its military modernization programme. However, China has bigger strategic fish to fry, mainly with respect to the United States and Japan. Besides, while the India-China relationship has its share of tensions, it is in practice a mix of common and conflicting interests and the former has yielded a steadily expanding trade relationship, making China India's largest trading partner today. As the level of interaction grows, the incentives for cooperation grow alongside, which raises expectations of rising interdependence and cooperation.

The domestic policy model focuses on internal change.

That seems unlikely. So far, the more right-leaning BJP has taken the critical step of declaring India's capability for and adherence to nuclear deterrence, but has not displayed an interest in rapid nuclearisation or in a more visible posture. More extremist parties are unlikely to come to power in what is an increasingly fragmented political system characterized by multiparty coalitions at the central as well as the state levels.

Such a system has the capacity to hinder drastic action rather than facilitate it. Nor does it seem reasonable to expect that domestic lobbies will have any more influence on nuclear strategy than they have had hitherto. Finally, with neither external nor domestic factors likely to impinge significantly on India's security, there is little cause for norms to change, in which case we can expect to see the norms against a hyper-active nuclear policy prevail. While the news media have a tendency to indulge in "hype" about India as a budding superpower, the leadership remains pragmatic and cautious (Gentleman, 2006). The latter sees the world as one characterized by increasing interdependence and hence requiring growing cooperation. [22] Looking back at two decades of experience with the bomb, and the decades of restraint prior to that, we can say with some confidence that together they have produced a trajectory that points toward continuing nuclear prudence in times to come.

The utility of this analysis extends beyond the central question about India's nuclear-strategic future. The accurate assessment of nuclear motivation requires a more complex view of the three models than has been the case. As I have shown, each model has components that may come to the fore at different times or be in tension with one another at any given time. In the present case, since the trajectories of each tend to coincide, a summation has not been problematic. But should they point in different directions, the overall picture may require a more careful analysis.

Whichever be the case, the approach should be useful in understanding the motives of existing and future nuclear powers while assessing strategic futures and policies geared to deal with them. Two types of questions to which this approach may be applied are, first, how will new nuclear powers such as North Korea and Iran think and act in the years ahead; and, second, how will the various nuclear-armed states respond to the call for universal disarmament? It is hoped that, building on this paper, future analyses will be able to address these critical questions better.

End Notes

1. There are innumerable works elaborating this point, but the classic remains Kenneth N. Waltz, *Theory of International Politics* (Reading, MA: Addison-Wesley, 1979).
2. On personality types and nuclearisation decisions, see Jacques E. C. Hymans, *The Psychology of Nuclear Proliferation* (Cambridge: Cambridge University Press, 2006).
3. On the concept of strategic culture, see Alastair Iain Johnston, *Cultural Realism: Culture and Grand Strategy in Chinese History* (Princeton, NJ: Princeton University Press, 1995). See also Jeffrey S. Lantis, "Strategic Culture and National Security Policy," *International Studies Review*, 4, 3 (Fall 2002), pp. 87-113.
4. I draw upon just one of three methodological pathways – the funnel strategy–identified by the authors as the other two are not relevant here. James Mahoney and Richard Snyder, "Rethinking Agency and Structure in the Study of Regime Change," *Studies in Comparative International Development*, Vol. 34, No. 2 (Summer 1999), pp. 3-32.
5. On strategic autonomy and the utility of nuclear weapons as a "backstop" to foreign policy, see Muthiah Alagappa, "Exploring Roles, Strategies and Implications: Historical and Conceptual Perspectives," in Muthiah Alagappa, ed., *The Long Shadow: Nuclear Weapons and Security in 21st Century Asia* (Stanford, CA: Stanford University Press, 2008), p. 85.
6. For the argument that the decision to test was primarily the result of domestic political considerations, see Kanti Bajpai, "The BJP and the Bomb," in Scott D. Sagan, ed. Inside Nuclear South Asia (Stanford, CA: Stanford University Press, 2009); Haider Nizamani, *The Roots of Rhetoric: Politics of Nuclear*

Weapons in India and Pakistan (Westport, CT and London: Praeger, 2000); Kalpana Sharma, "The Hindu Bomb," *Bulletin of the Atomic Scientists*, 54 (4) (July-August 1998), pp. 30-33.

7. The original concept held that nuclear stability permits conventional war. But that has never really happened. Instead, nuclear rivalries have been characterized by proxy conflict and marginal fighting. See Glen Snyder (1965): "The Balance of Power and the Balance of Terror," in Paul Seabury, (ed.), *The Balance of Power*, San Francisco: Chandler. For a discussion of the phenomenon in the South Asian context, see Krepon, Michael and Chris Gagné, (eds.), *The Stability-Instability Paradox: Nuclear Weapons and Brinkmanship in South Asia*, Henry L. Stimson Center, Washington, DC, June 2001.
8. Cited in "Impact of a Nuclear Strike," *BBC News*, May 29, 2002 http://news.bbc.co.uk/2/hi/south_asia/2012543.stm> (accessed 29 April 2009).
9. The domestic politics of the initial phase, with the government's weakness becoming manifest early, is discussed in Dinshaw Mistry, "Diplomacy, Domestic Politics, and the US-India Nuclear Agreement," *Asian Survey*, 46 (5) (September/October 2006), pp. 675-698.
10. The debate has yet to receive adequate scholarly attention, but a perusal of the reports of the debate between 2005 and 2008 show the extensive role played by the Department of Atomic Energy (DAE) both in the media debate and in the actual negotiations with the United States.
11. "Uranium Holds Key to Energy Security," *Hindu*, February 6, 2009 <http://www.thehindu.com/2009/02/06/stories/2009020660401200.htm> (accessed February 6, 2009).
12. On the civil-military bargain, see Sunil Dasgupta, "The Indian Army and the Problem of Military Change," in Swarna Rajagopalan, ed., *Security and South Asia: Ideas, Institutions, and Initiatives* (London and New Delhi: Routledge, 2006).
13. Government of India, *Reforming the National Security System: Recommendations of the Group of Ministers*, February 2001.
14. "The Cabinet Committee on Security Reviews Operationalisation of India's Nuclear Doctrine," Press Release, Ministry of External Affairs, Government of India, January 4, 2003<

http://meadev.nic.in/news/official/20030104/official.htm> (accessed on February 19, 2003).

15. For a critique of this fundamental flaw in realist thought, see Campbell Craig, *Glimmer of A New Leviathan: Total War in the Realism of Niebuhr, Morgenthau, and Waltz* (New York: Columbia University Press, 2003).
16. Marc Trachtenberg, "The Influence of Nuclear Weapons in the Cuban Missile Crisis," *International Security*, Vol. 10, No. 1 (Summer 1985), pp. 137-163. Comparative studies of nuclear rivalries have been few. For an exception (covering Cold War era conflicts) that supports my point here, see Rajesh Rajagopalan, *Second Strike: Arguments about Nuclear War in South Asia* (New Delhi: Viking, 2005), pp. 89-106.
17. Representative writings include Gurmeet Kanwal, *Nuclear Defence: Shaping the Arsenal* (New Delhi: Knowledge World, 2001); Bharat Karnad, *Nuclear Weapons and Indian Security: the Realist Foundations of Strategy* (Delhi: Macmillan, 2002); and Raja Menon, *A Nuclear Strategy for India* (New Delhi, Thousand Oaks & London: Sage, 2000).
18. Government of India, Ministry of External Affairs, *Draft Report of National Security Advisory Board on Indian Nuclear Doctrine*, August 17, 1999 < http://meaindia.nic.in/> (accessed May 1, 2009).
19. *Indian Express*, Pranab Defends Foreign Policy, Says Attack No Option," February 19, 2009, <http://www.indianexpress.com/news/pranab-defends-foreign-policy-says-attack-no-option/425301/> (accessed February 19, 2009).
20. *Dawn*, "Diplomacy, Not Bullet Power, Worked: Mukherjee," April 17, 2009 <http://www.dawn.com/wps/wcm/connect/dawn-content library/dawn/news/world/diplomacy-not-bullet-power-worked-mukherjee--bi> (accessed April 17, 2009).
21. *Global Security Newswire*, "Pakistani Nuke Arsenal on Track to Be World's Fifth Largest," February 1, 201, available at, <http://gsn.nti.org/gsn/nw_20110201_5282.php>(accessed same day).
22. See for example Prime Minister Manmohan Singh's views in "'Inadequate Recognition of Growing Interdependence,'" *Hindu*, November 7, 2006, <http://www.thehindu.com/2006/11/07/stories/20061107156212

00.htm> (accessed May 2, 2009). See also "India as a Rising Great Power: Challenges and Opportunities," Speech by M. K. Narayanan, India's National Security Advisor, at the IISS-Citi India Global Forum, April 18-20, 2008, <http://www.iiss.org/conferences/iiss-citi-india-global-forum/igf-plenary-sessions-2008/fourth-plenary-session-india-and-the-great-powers/fourth-plenary-session-m-k-narayanan/> (accessed April 25, 2008).

References

Abraham, Itty (1998): *The Making of the Indian Atomic Bomb*, New Delhi: Orient Longman.

Abraham, Itty (1997): "Science and Secrecy in Making of Colonial State," *Economic and Political Weekly,* pp. 2136-2146.

Alagappa, Muthiah (2008): "Exploring Roles, Strategies and Implications: Historical and Conceptual Perspectives," in Muthiah Alagappa, (ed.), *The Long Shadow: Nuclear Weapons and Security in 21st Century Asia*, Stanford, CA: Stanford University Press, p. 85.

Anand, Y.P. (1998): *What Mahatma Gandhi Said about the Bomb*, New Delhi: National Gandhi Museum.

Baru, Sanjaya (2006): *Strategic Consequences of India's Economic Performance*, New Delhi: Academic Foundation.

Basrur, Rajesh M. (2008): "India: Imbalance under Civilian Control," in Thomas Bruneau and Harold Trinkunas, Eds., *Global Politics of Defense Reform,* New York: Palgrave Macmillan.

Basrur, Rajesh M. (March 2001): "Nuclear Weapons and Indian Strategic Culture," *Journal of Peace Research*, 38 (2): 181-198.

Basrur, Rajesh, Timothy Hoyt, Rifaat Hussain and Sujoyini Mandal (2009): *The 2008 Mumbai Terrorist Attacks: Strategic Fallout*, Singapore: S. Rajaratnam School of International Studies.

Booth, Ken and Nicholas Wheeler, (2008): *The Security Dilemma: Fear, Cooperation and Trust in World Politics*, Houndmills, Basingstoke: Palgrave Macmillan.

Center for Non-proliferation Studies, Monterey Institute of International Studies, (CNPS) (1999): "China's Nuclear Exports and Assistance to Pakistan," Monterey, CA, August <http://cns.miis.edu/research/india/china/npakpos.htm> (accessed April 30, 2008).

Chari, P.R. Pervaiz Iqbal Cheema and Stephen P. Cohen, (2003): *Perception, Politics and Security in South Asia: The Compound Crisis of 1990*, London: Routledge Curzon, pp. 139-140, 135.

Chari, P.R., Pervaiz Iqbal Cheema and Stephen P. Cohen, (2007): *Four Crises and A Peace Process*, Washington DC: Brookings Institution Press.

Chengappa, Raj (2000): *Weapons of Peace*, New Delhi: Harper Collins, pp. 88-89, 332-6.

Cohen, Stephen P. 2000): "Why Did India 'Go Nuclear'?" in Raju G. C. Thomas and Amit Gupta, eds., *India's Nuclear Security*, New Delhi: Vistaar Publications.

Craig, Campbell (2003): *Glimmer of A New Leviathan: Total War in the Realism of Niebuhr, Morgenthau, and Waltz*, New York: Columbia University Press.

Dasgupta, Sunil (2006): "The Indian Army and the Problem of Military Change," in Swarna Rajagopalan, Ed., *Security and South Asia: Ideas, Institutions, and Initiatives*, New Delhi: Routledge.

Dawn (2009): "Diplomacy, Not Bullet Power, Worked: Mukherjee," April 17, <http://www.dawn.com/wps/wcm/connect/dawn-content-library/dawn/news/world/diplomacy-not-bullet-power-worked-mukherjee--bi> (accessed April 17, 2009).

Devotta, Neil (2003): "Is India Over-extended? When Domestic Disorder Precludes Intervention", *Contemporary South Asia*, 12 (3): 365-380.

Elias, Norbert (1991): *The Symbol Theory*, New Delhi: Sage Publications, pp. 57-58, 72.

Fitzpatrick, Mark (May 2008): "US-India Nuclear Cooperation Accord: Implications for the Non-proliferation Regime," *Asia-Pacific Review*, 15 (1): 76-85.

Frey, Karsten (2002): *State Interests and Symbolism in India's Nuclear Build-up*, South Asia Institute, Department of Political Science, University of Heidelberg.

Ganguly, Sumit (2000): "Explaining the Indian Nuclear Tests of 1998," in Raju G., C. Thomas and Amit Gupta, (eds.), *India's Nuclear Security*, New Delhi: Vistar Publications.

Ganguly, Sumit (2008): "Nuclear Stability in South Asia," *International Security*, 33 (2): 45-70.

Ganguly, Sumit and Devin T. Hagerty (2005): *Fearful Symmetry: India-Pakistan Crises in the Shadow of Nuclear Weapons*, New

Delhi: Oxford University Press.

Gentleman, Amelia (2006): "India Can't Wait to Put the 'Super' before 'Power,' " *International Herald Tribune*, November 23, available at, <http://www.defence.pk/forums/india-defence/2954-letter-india-india-cant-wait-put-super-before-power.html> (accessed May 1, 2009).

Gilpin, Robert (1981): *War and Change in World Politics*, Cambridge: Cambridge University Press, pp. 30-33.

Giridharadas, Anand (2008): "Land of Gandhi Asserts Itself as Global Military Power," *New York Times*, September 22, <http://www.nytimes.com/2008/09/22/world/asia/22india.html?ref=world> (accessed September 22, 2008).

Glaser, Charles L. (1997): "The Security Dilemma Revisited," *World Politics*, 50 (1): 171-201.

Global Security Newswire (2001): "Pakistani Nuke Arsenal on Track to Be World's Fifth Largest," February 1, available at, <http://gsn.nti.org/gsn/nw_20110201_5282.php>(accessed same day).

Gourevitch, Peter A. (2002): "Domestic Politics and International Relations," in Walter Carlsnaes, Thomas Risse, and Beth Simmons, (eds.), *Handbook of International Relations* London: Sage Pub.

Hagerty, Devin T. (1999): "South Asia's Big Bangs: Causes, Consequences, Prospects," *Australian Journal of International Affairs*, 53 (1): 19-29.

Hymans, Jacques E.C. (2006): *The Psychology of Nuclear Proliferation: Identity, Emotions, and Foreign Policy,* Cambridge: Cambridge University Press.

Indian Express (2009): Pranab Defends Foreign Policy, Says Attack No Option," February 19, <http://www.indianexpress.com/news/pranab-defends-foreign-policy-says-attack-no-option/425301/> (accessed February 19, 2009).

Jervis, Robert (1978): "Cooperation under the Security Dilemma," *World Politics*, 30(2): 1167-1214.

Jervis, Robert (1989): *The Meaning of the Nuclear Revolution: Statecraft and the Prospect of Armageddon,* London: Cornell University Press, pp. 174-225.

Jervis, Tom (1989): *Meaning of the Nuclear Revolution*; *Statecraft and the prospect of Armageddon,* New York: Cornell University Press, pp.180-185.

Jha, Prem Shanker (1998): "Why India Went Nuclear," *World Affairs*, 2(3): 80-96.

Johnston, Alastair Iain (1995): *Cultural Realism: Culture and Grand Strategy in Chinese History,* Princeton, NJ: Princeton University Press.

Kanwal, Gurmeet (2001): *Nuclear Defence: Shaping the Arsenal,* New Delhi: Knowledge World.

Kanwisher, Nancy (1989): "Cognitive Heuristics and American Security Policy", *Journal of Conflict Resolution*, 33 (4): 652-675.

Kapur, Ashok (1976): *India's Nuclear Option*, New York: Praeger, pp. 193-194.

Kapur, Paul (2008): "Ten Years of Instability in a Nuclear South Asia," *International Security*, 33 (2): 71-94.

Kapur, Paul S. (2007): *Dangerous Deterrent: Nuclear Weapons Proliferation and Conflict in South Asia* Stanford, CA: Stanford University Press.

Kapur, Paul S. (2007): *Dangerous Deterrent: Nuclear Weapons Proliferation and Conflict in South Asia*, Stanford, CA: Stanford University Press.

Karnad, Bharat (2002): *Nuclear Weapons and Indian Security: the Realist Foundations of Strategy* Delhi: Macmillan.

Kraig, Michael Ryan (2003): "The Political and Strategic Imperatives of Nuclear Deterrence in South Asia," *India Review,* 2 (1): 30.

Krepon, Michael (2008): "Looking Back: the 1998 Indian and Pakistani Nuclear Tests," *Arms Control Today,* http://www.armscontrol.org/act/2008_05/lookingback.asp>(accessed May 10, 2008).

Krepon, Michael and Chris Gagné, (eds.), (2001): *The Stability-Instability Paradox: Nuclear Weapons and Brinkmanship in South Asia*, Washington, DC.: Henry L. Stimson Center.

Ladwig III, Walter C. (2007/2008): "A Cold Start for Hot Wars?", *International Security*, 32 (3): 158-190.

Lantis, Jeffrey S. (2002): "Strategic Culture and National Security Policy," *International Studies Review*, 4 (3): 87-113.

Mahoney, James and Richard Snyder (1999): "Rethinking Agency and Structure in the Study of Regime Change," *Studies in*

Comparative International Development, 34 (2): 3-32.

Mehta, Pratap Bhanu (1998): "India: the Nuclear Politics of Self-Esteem," *Current History*, 97, 623: 403-406.

Menon, Raja (2000): *A Nuclear Strategy for India,* New Delhi, Thousand Oaks and London: Sage.

Ministry of External Affairs (1999): *Draft Report of National Security Advisory Board on Indian Nuclear Doctrine*, Government of India, August 17, < http://meaindia.nic.in/> (accessed May 1, 2009).

Mistry, Dinshaw (2006): "Diplomacy, Domestic Politics, and the US-India Nuclear Agreement", *Asian Survey*, 46 (5): 675-698.

Mitra, Subrata K. (2003): "The Reluctant Hegemon: India's Self-perception and the South Asian Strategic Environment", *Contemporary South Asia,* 12 (3).

Morgan, Patrick M. (2003): *Deterrence Now: A Conceptual Analysis*, Beverly Hills, CA: Sage Publications, pp. 80-86.

Nizamani, Haider (2000): *The Roots of Rhetoric: Politics of Nuclear Weapons in India and Pakistan,* London: Praeger.

O'Neill, Barry (1999): *Honor, Symbols and War*, Ann Arbor: University of Michigan Press.

Ollapally, Deepa (2001): "Mixed Motives in India's Search for Nuclear Status," *Asian Survey*, 41(6): 925-942.

Pant, Harsh V. (2007): "India in the Asia-Pacific: Rising Ambitions with an Eye on China," *Asia-Pacific Review*, 14 (1): pp. 54-71.

Paul, T.V. (2007): "The US-India Nuclear Accord: Implications for the Non-proliferation Regime", *International Journal*, 62 (4): 845-861.

Paul, T.V. (2003): "Chinese-Pakistani Missile Ties and the Balance of Power," *Non-proliferation Review*, 10 (2): 1-9.

Perkovich, George (1999): *India's Nuclear Bomb: The Impact on Global Proliferation,* Berkeley, CA: University of California Press.

Pubby, Manu (2009): "First External Review Lists Process to revamp DRDO," *Indian Express*, March 2, <http://www.indianexpress.com/news/first-external-review-lists-process-to-revamp-drdo/429607/>(accessed March 2, 2009).

Rabasa, Angela et al., (2009): *The Lessons of Mumbai*, Santa Monica, CA: RAND.

Rajagopalan, Rajesh (2005): *Second Strike: Arguments about*

Nuclear War in South Asia, New Delhi: Viking, pp. 89-106.

Riedel, Bruce (2008): "South Asia's Nuclear Decade," *Survival*, 50 (2): 107-126.

Sagan, Scott D. (1996/97): "Why Do States Build Nuclear Weapons? Three Models in Search of a Bomb," *International Security*, 21 (3): 54-86.

Sagan, Scott D. (ed.) (2009): *Inside Nuclear South Asia*, Stanford, CA: Stanford University Press.

Saksena, Jyoti (2006): "Regime Design Matters: The CTBT and India's Nuclear Dilemma," *Comparative Strategy*, 25 (2): 209-229.

Scott, David (2007): "Strategic Imperatives of India as an Emerging Player in Pacific Asia", *International Studies*, 44 (2): 123-140.

Sharma, Dhirendra (1983): *India's Nuclear Estate*, New Delhi: Lancer International.

Sharma, Kalpana (1998): "The Hindu Bomb", *Bulletin of the Atomic Scientists*, 54 (4): 30-33.

Snyder, Glen (1965): "The Balance of Power and the Balance of Terror," in Paul Seabury, (ed.), *The Balance of Power*, San Francisco: Chandler.

Solingen, Etel (2007): *Nuclear Logics: Contrasting Paths in East Asia and the Middle East*, Princeton and Oxford: Princeton University Press.

Solingen, Etel (2007): *Nuclear Logics: Contrasting Paths in East Asia and the Middle East*, Princeton, NJ: Princeton University Press.

Subrahmanyam, K. (1998): "Indian Nuclear Policy, 1964-98 (A Personal Recollection)", in Jasjit Singh, (ed.), *Nuclear India*, New Delhi: Knowledge World, p. 44.

Subrahmanyam, K. (2004): "Politics of Security: When Vajpayee Said 'No' to Going Nuclear," *Times of India*, April 10, <http://timesofindia.indiatimes.com/articleshow/608713.cms> (accessed March 16, 2006). Vajpayee's objection was prompted by the desire not to provoke Pakistan.

Subrahmanyam, K. "Nuclear India in Global Politics," *World Affairs*, 2 (3): 12-40.

The Hindu (1999): "India Not to Engage in a Nuclear Arms Race: Jaswant," (Interview), November 29, p. 14.

The Hindu (1999): "India Not to Engage in a Nuclear Arms Race: Jaswant," (Interview), November 29, p. 14.

The Hindu (1999): "India Not to Engage in a Nuclear Arms Race: Jaswant," (Interview), November 29, p. 14.

The Hindu (2009): "Uranium Holds Key to Energy Security," February 6, <http://www.thehindu.com/2009/02/06/stories/2009020660401200.htm> (accessed February 6, 2009).

Trachtenberg, Marc (1985): "The Influence of Nuclear Weapons in the Cuban Missile Crisis," *International Security*, 10 (1): 137-163. Comparative studies of nuclear rivalries have been few.

Weiss, Leonard (2007): "US-India Nuclear Cooperation," *Non-proliferation Review*, 14 (3): 429-457.

13

Redefining Regional Co-operation in South Asia

Anjali Sharma

Geographically, South Asia is a region which may either be seen as a collection of smaller units or as one part of a larger whole. With the core seven countries, now eight, viz. Afghanistan, Bangladesh, Bhutan, India, Maldives, Nepal, Pakistan, Sri Lanka, the region covers 2.4 percent of the world's total land surface area and accounts for over 16.5 percent of the world's total population. For almost two hundred years in the millennia old history of South Asia, it was under the British rule except Nepal and Bhutan which served more or less as the British protectorates. Unable to manage the far-off colonies after the end of the World War II in 1945 that completely drained out the energies and economies of the European nations, British Empire gradually started the process of granting independence to most of them.

In South Asia, Pakistan and India got the independence in 1947 followed by Sri Lanka in February 1948. Maldives followed the suit in 1965. Internal disturbances within Pakistan led to the emergence of Bangladesh in 1971 and Sikkim merged with India in 1975, thus completing the cartographic process of the region building in South Asia.

While regional boundaries were more or less appeared to be settled by the year 1975, the long years of foreign occupation, dependence on foreign aid, fragile economy, unstable polity, population explosion and its associated maladies, corruption and nepotism delay the process of regional unification which should have been the natural corollary of the region building in South Asia. Had it been achieved on time, the process of regional cooperation in South Asia would have been a great success by now.

It is easy to identify the territorial boundaries of South Asia on map but it is equally difficult to find the true identity of the region. My paper aims to highlight this malady of regional cooperation in South Asia in the following paragraphs. I will also try to find out the suitability of regional organizations in a highly fragmented region of South Asia. Besides, an attempt will also be made to find out an answer to the future possibilities of a successful regional integration in South Asia. A brief perusal of the other successful regional organizations like Association of South East Asian Nations (ASEAN) and European Union (EU) is also made to compare their successes with the South Asian Association for Regional Cooperation (SAARC) failures.

The debate over the relevance and irrelevance of SAARC which is raging for quite some time has been confronted with the two extreme viewpoints. The first group comprise of eternal optimists argues that SAARC albeit its slow process is a living phenomena and thus, cannot be wished away. They are hopeful that in few years SAARC will be kicking and thriving just like EU which too takes nearly 50 years to take off. They quote the then Indian foreign minister Pranab Mukherjee's statement in their support. During the 14th SAARC summit in 2007, Mukherjee told the media that "SAARC is an idea whose time has come.

The cooperation and accommodation that have characterized our negotiations and deliberations show our countries are ready for a transformation of SAARC into an effective instrument of regional advancement" (Guha, 2007). However, the second group of analysts put question mark on the achievements of SAARC. They point out and rightly so that every attempt at regional cooperation has been made futile by Pakistan directly and India indirectly. These two countries are to be blamed for the failure of SAARC which merely provide luxurious getaways and photo opportunities to the leaders of the region. The sceptics use the words 'SAARC' and 'NATO' interchangeably, which means 'No Action Talk Only'.

While both the groups have raised valid points related to the purposeful existence of SAARC, a moderate viewpoint is needed to reconcile the two extremes together. Without finding a middle path, regional cooperation in South Asia will go haywire which cannot be considered as the best option in a globalised world. A contemporary world is also an interdependent world where one has to learn to use the scarce resources in a best possible manner. It is a world where the resources and services are unevenly divided among the different territorial boundaries. As no country is self-sufficient, they cannot survive in isolation.

This necessitates cooperation which involves joint action and collaboration in order to achieve a shared advantage. While whole world is now globalized and cooperating in one way or other, the most appropriate and convenient form of cooperation is the regional cooperation. As the region is a geographically contiguous area, it is easier to cooperate in different sectors and share fruits of the mutual joint ventures at the same time. For example, transport and communication connectivity can be most successful and better achieved within the region. Same is the case with infrastructure development, issues like trade and environment where regional initiatives rather then international networking, matters most.

Such realization has still not dawned upon the South Asian policy making circles. Or, may be who knows that the realization is very much present but the political will as an important pre-condition for regional cooperation to give it a concrete shape is conspicuous by its absence. The reasons may be many but the fact remains that regional cooperation in South Asia has so far failed to make its impact felt in resolving many issues of regional importance. It is because very foundation on which the leaders of South Asia erected the structure of regional cooperation is ill-thought out.

Theoretical Basis of Regional Cooperation in South Asia

The need for regional cooperation was felt first by the smaller countries of South Asia, mainly Bangladesh. In fact, the concept of SAARC was first mooted by the late President Mujib-ur-Rehman of Bangladesh in 1975. It took several rounds of preparatory meetings for the SAARC to get into a formal shape in 1985. It was planned to be an annual affair but soon bilateral disputes, misunderstandings and mistrust between neighbours, especially between India and Pakistan began to take its toll resulted in the postponements of a number of summits often at the behest of India. India's ability to call shots in the region at will fuelled further suspicion in the minds of its neighbours regarding its status as being a 'mini-imperialist power' in South Asia.

While smaller countries of the region started viewing India as some form of a 'bully', India too was not much impressed by the concept of regional cooperation in South Asia which it thought was some kind of a 'ganging up' of small countries against its might. Such a negative attitude displayed by its members from the beginning of the organization itself failed to give SAARC a much-needed jump start which could transform it into a truly regional body. These initial hiccups and weaknesses at birth made SAARC a limping and impotent body unable to produce tangible results.

Apart from the general causes which are responsible for the failure of the SAARC, there were certain theoretical issues which South Asian policy makers failed to take into account while formulating SAARC. The most important question which they failed to address during their deliberations was that whether South Asia is really prepared to embrace the advanced concept of regional cooperation? Nobody tested the waters before stepping into the swimming pool.

Given the history of one of the worst partitions in world history, it is quite surprising that how the leaders have failed to take into account the preparedness of the region as far as regional cooperation is concerned. SAARC was formed as a political body without taking general public into confidence.

No efforts were made to popularize the concept among the local population. People came to know of it much later when SAARC failed to succeed in its ventures owing to India-Pakistan rivalry. Inability of the leaders of the South Asian countries to establish a relationship between the institutional sphere of society and the inhabitants of the region actually failed the South Asian Regional Organization as this is the relationship which, according to the Finnish Geographer Ansi Passi, forms the 'fundamental meeting place where social consciousness and as part of the latter, regional consciousness is continually produced' (Paasi and Knapp, 2003).

I would like to delve more deep to understand the relationship between the region and its inhabitants and its impact on the success and failure of regional cooperation. In the words of Ansi Passi 'a region is mediated in our everyday life in the form of various symbols, which are the same for all individuals in the one region—though the meanings associated with them will always be construed personally on the basis of individual's life, situation and geography'(Paasi and Knapp, 2003). When the process of regional cooperation was started first in 1975, South Asia was not accepted as a region by its inhabitants as well as leaders who find it more convenient to raise the bogey of 'nationalism under threat' to secure their vote banks rather then getting into the nuances of regional cooperation.

As a result, persons who were entrusted with the responsibility of putting regional cooperation into shape have never accepted the region as a historically continuous process which could have facilitated an understanding of a specific regional identity (Paasi and Knapp, 2003). The historical facts were brutally distorted to safeguard the personal interests of those in power blurring the fine lines in South Asian history which obstructed the emergence of a common regional identity. There were no common images or symbols or motto (except cultural bonds which are too personal to unite a region) that could bind the eight countries of the region together into a single whole. This resulted into the failure of regional cooperation in South Asia.

Ansi Passi in his paper 'A Sense of Place' has identified four different aspects of the historical formation of concrete regions in a society. I am applying this classification to the region of South Asia to measure the extent to which the regional cooperation can succeed in the region. The first is the *assumption of the territorial shape*. South Asia has a definite shape and place in the world's map. Its boundaries are clearly defined and properly defended and there is no question of their redrawing under any circumstances. The boundary question involving the long history of colonialism and independence accompanied by the large scale bloodshed was settled long ago. There is no denying the fact that geographically, South Asia is a region.

The second factor hinges on the *development of conceptual (symbolic) shape of the region.* The region of South Asia failed to measure itself up to this criterion as the inhabitants would like to identify themselves on a national basis. The people do not find any common ground between their country and the region despite the fact that they are confronted with the similar problems and issues which needs common solutions in their daily lives.

Strong national identities have always relegated the question of regional identity in South Asia into a background. Notwithstanding the apathy of the regional population which is largely apolitical, South Asia doesn't have a leader with a regional vision who can also be a statesman. India too, has dashed all expectations in this regard when it appeared as a country which is already pre-occupied with the South Asian power games and as a result, has failed to display the vision for South Asia.

Third factor is the *development of the sphere of institutions* where South Asia fares moderately. A number of regional institutions including SAARC have been created to foster the spirit of regional cooperation in the region yet not all of them have attained their full potential. Majority of them got mired into the vortex of regional politics and soon faded into oblivion. It is however, worthy to note that SAARC so far remained more or less a permanent feature of South Asia. In fact, SAARC is the only regional organization which gives the region of South Asia its permanent identity. SAARC represents South Asia and it is through this window that the world corresponds with South Asia.

The fourth factor is the *development of regional consciousness of a society* which is gradually gaining ground in South Asia. This is happening due to the rapid advances in the means of telecommunication in South Asia which enables people to interact and share their problems and solutions in virtual world despite national restrictions on movement across borders. Soft power is emerging as a potent source for advancing regional brotherhood sans boundaries in place of hard power which these days has find its use severely restricted to border areas alone.

The order of elements/factors mentioned above may differ depending upon the regional specifications but nevertheless, they are indispensable for creating concrete regions in a society. A brief analysis finds South Asia at different stages of development vis-à-vis the above-mentioned factors which may be interpreted further to assess the future of regional cooperation in South Asia.

Expectations and Regional Identity in South Asia

The 'structures of expectations' which Ansi Passi considered as essential to generate the feeling of regional identity among the inhabitants of the region depends upon the nature of one's experiences of the world or the region in a given culture. If the nature of the experiences gained by the inhabitants of the region is bad, then such structures will fail to create common regional identity in any region, not just in South Asia.

On the contrary, good experiences will always help create a strong bonding of the people with their region. The concept of Structures of Expectations which is usually employed in linguistic and cultural arenas depicts the ways in which people organize their knowledge of the world and use it to generate interpretations and relationships with regard to new information and experiences. These structures refer to both the physical and cultural character of a region and can comprise real, imagined and mythical features. In fact, they are essential for establishing the past and the future as a common frame of reference for the inhabitants of a region.

In South Asia, such structures of expectations often present in the form of history covering colonial domination followed by the bloody partition, cross-cultural linkages shaped by religion, disputed territories as historical leftovers, size disparities and uneven stages of development of the member countries are not at all conducive for uniting the region as a whole. The attempt was however, made to create a common Structures of Expectations through SAARC in 1985 when 11 common issues were identified for mutual regional cooperation but it was too little, too late. Although there is no denying the fact that the destiny of South Asia is intertwined, its realization began gradually and incrementally.

Regional Integration versus Regional Co-operation

Regional integration can be defined along three dimensions:

1. Geographic scope illustrating the number of countries involved in an arrangement.
2. The substantive coverage or the width that is the sector or the activity concerned.
3. The depth of integration to measure the degree of sovereignty a country is ready to surrender, that is from simple coordination or cooperation to deep integration (Kritzinger and Niekerk).

Keeping in view the applicability of the above-mentioned

factors to facilitate regional integration in South Asia, it should be worth considering that with the addition of Afghanistan as the eighth member of SAARC, the geographic scope or the variable geometry of South Asia is almost complete. There is also more or less an agreement on the scope of activities to be brought under the regional umbrella. It is noteworthy that the list of activities to be included under the rubric of SAARC is gradually increasing. However, the countries have differences on the issue concerning the depth of integration. Therein lays the problem which poses challenges to the regional integration of South Asia. As South Asia comprise of the countries which are fiercely nationalist, the question of even minimal surrender of their national sovereignty for the sake of regional integration is an anathema to them.

While complete regional integration, the kind of which is present in Europe and South East Asia in the form of European Union and Association of South East Asian Nations respectively is hard to achieve in South Asia, at least in the near hundred years or so, its lesser forms viz. regional cooperation and coordination based on the compatibility of different member countries with each other can certainly be think of as the good alternatives to full-fledged integration.

Given the fact that all the member countries of SAARC are at the different stages of development, it is futile to think of anything even remotely related with the South Asian Economic Union or South Asian Customs Union. South Asian common currency, the proposal of which is in the offing for quite some time now, is also a vague concept because of the fragility and uncertainty of the existing economic and political institutions of South Asia. Standardization is needed for the completion of the process of regional integration anywhere in the world and South Asia is no exception. Certain eligibility criteria should be set before embarking on the process of regional integration.

Total regional integration which the analysts earlier thought of as the panacea of all ills is no more an attractive proposition, especially in the region as diverse as South Asia. The experience of the European Union is there for all of us to see and learn lessons from it. Despite setting certain requirements for its members, EU is currently confronting with the task of managing the economic fallout resulting from the Greece financial crisis. Portugal, Spain and many other countries of the EU are at the verge of bankruptcy, a situation which will also impact their common currency Euro. This proves that complete regional integration is not fully successful even in the most successful and stable regions of the world.

On the other hand, the concepts of regional cooperation and harmonization/coordination are worth applying in the setting of South Asia where historically, animosities prevail between member countries. Regional cooperation may be the weakest and issue-focused arrangement where countries may cooperate on various joint development projects and broad issues with full sovereign control over the 'right to exit'. While cooperation can be said to be the lowest level of multilateral commitment, harmonization or coordination implies a higher and more formalized degree of cooperation and commitment from the member countries. Cooperation and harmonization/coordination processes do not need a decision making supranational body, although a higher degree of joint commitment is needed (Kritzinger and Niekerk). The two concepts can be worth trying before going for complete regional integration to create trust and understanding between the different stake holders in South Asia.

Pattern of Regional Co-operation in South Asia

Regional cooperation in South Asia is based on an occasional regional get together. Annual meetings of senior leaders of various South Asian countries provide stimulus to various regional projects which would have been left to fend for themselves, had there been no meetings between the Heads of the member countries under the rubric of SAARC. In this way SAARC, despite being a lame duck association has been doing a greatest service to its more then a billion people whom the body claims to represent. Annual get together which has earned the sobriquet of a 'Talk Shop' for SAARC has led to the signing of various agreements although the number of and actual functioning of such agreements are far below expectations. But one has to accept the fact that these agreements are by and large the product of the SAARC summits.

In South Asia, Regional Integration and Regional Coordination which are the by-products of Regional Cooperation cannot be said to be supplementing each other's efforts. While South Asian University (SAU) can be considered as the step forward towards regional integration in the troubled region, regional coordination and harmonization is severely lacking even in this noble initiative. Other countries are not in agreement with India over the recruitment and the pay scales of the university teachers which they found as exorbitantly high in India. There are fewer students from Pakistan in its first batch and no official efforts have been made to popularize the concept of the regional university among the South Asian population. Such instances can be further multiplied.

Regional Cooperation in South Asia is taking place within the occasional get together of the regional leaders under the banner of SAARC annual summits. Regional Integration and Regional Coordination despite being the offshoots of Regional Cooperation is working independently of each other which is one of the causes for the failure of regional initiatives in the region.

Niekerk, a senior economist with the World Bank in his paper 'Regional Integration: Concepts, Advantages, Disadvantages and Lessons of Experience' has mentioned some key principles considered to be indispensable for the success of any initiative taken towards regional integration. All the three principles mentioned below are not present in South

Asia but nonetheless are workable given the real political will to materialize the so far utopian concept of Regional Integration in South Asia.

Open Regionalism: The concept has its origin in Asia Pacific Economic Cooperation (APEC) which is based on the idea that member states liberalizes intra-bloc trade while at the same time lowering external trade barriers on imports from the rest of the world. The concept denotes a region-wide market economy that is also opened to the outside world. In South Asia, almost all economic initiatives like South Asian Free Trade Area (SAFTA) and South Asian Preferential Trade Agreement (SAPTA) ended in a failure. India, the largest economy in the region has bilateral free trade agreement only with one country i.e. Sri Lanka. Despite the fact that all member countries of SAARC have granted Most Favoured Nation (MFN) status to each other, Pakistan have not granted the same to India due to their cold ties. Regional market integration appears to be a far cry in South Asia due to such irritants.

According to Niekerk, Open Regionalism is central for assuring consistency and complementarities between national reform programmes and the sub-regional agenda. Open Regionalism involves three factors:

1. **Outward Orientation:** This means that sub regional policies must contribute to lowering and eliminating obstacles to global trade and investment, including tariff and non-tariff barriers. Economic diversification and expansion of non-traditional exports remains the key to accelerate growth in the region. The time has come to eradicate high levels of protectionism which are still maintained in South Asia due to mutual mistrust and misunderstandings. SAARC is more inward-looking which is responsible for its decadence. In order to become a vibrant organization, SAARC should look out to reach the world and try to make an attempt to collectively bargain with other trade blocs. At present, the focus of the SAARC

countries is to deal unilaterally with other countries and blocs outside the region which in turn is benefiting only a part of the region and not the region as a whole. A start has to be made to make the SAARC more outward oriented which seems to be a remote possibility, as of now.

2. **Market-driven Integration Process:** Governments must not develop national monopolies as well as multinational monopolies at the sub-regional level. Instead, competition should be encouraged across borders. Countries should cooperate through a common regional body to develop common infrastructural services which have traditionally remain the domain of national monopolies (Kritzinger and Niekerk).

   SAARC should be given the powers to undertake such joint initiatives to enhance the levels of cooperation in South Asia. It may not materialize soon but at least efforts must be made to achieve the end.
3. **Private Sector Involvement:** People of the region should be the critical elements and governments and regional organizations only facilitators through appropriate choices and policies. Private operators have to be involved in the design and implementation of regional activities. It is necessary to change the perception that regional organizations are simply remote outgrowths of government bureaucracies. In the functioning of SAARC, private sector involvement is almost negligible. People of the region generally feel neglected and alienated from any of the regional initiatives decided upon by the leaders of the member-countries as they are not involved in any stage of such programmes. Absence of general support and a total lack of private involvement led to the weakness of SAARC. SAARC should adopt a top-down approach and better still, a bottom-up approach to make the organization more people-friendly and consequently, vibrant.

Subsidiarity: Niekerk's paper mentioned subsidiarity as the second most important principle for a successful regional integration. It is also the term made popular by the European Union. It means that regional institutions should be responsible only for those activities that are not better handled at the national level. In SAARC, although eleven areas of cooperation (16, now) are clearly highlighted [1], yet most of the time they fall into the trap of regional inertia which hinges on bilateral misgivings and 'big country-small country' syndrome. In SAARC, even if the member country is completely failed to deal with a particular thing at a national level, it is certain that the country will not discuss it at the regional forum. Achieving consensus on any single issue is difficult in SAARC.

Incrementalism, Pragmatism and Standardization: Niekerk suggests that a credible integration must be built on pragmatic, gradual steps that reinforce trust and commitment, and make the process self-perpetuating. A clear vision is necessary to proceed forward towards the goal of regional integration and the movement should be gradual and incremental with timetables and targets that are credible and realistic. Proper care should be taken to see that the targets are met because unfulfilled targets and declaration leads to frustration and reversals. In addition, a certain workable criteria or standards should be set for every country of the region to become a member of the regional organization.

Membership of the European Union is based on the fulfilment of certain standards which plays a major role in its success. Admission of loose links (weak members) in the chain often led to disastrous results like Greece in the EU. Recent financial crisis engulfing Greece is responsible not only for devaluing Euro in the world market but also affected the economies of other smaller European nations like Portugal and Spain. From its inception, SAARC has not set any standard for its member countries to fulfil which put hindrances in achieving the goals set up by the regional body for itself. Although SAARC is taking incremental steps towards its goals, the process is not gradual but slow.

The pragmatic vision is absent and no standards are followed putting a question mark on the relevance of SAARC and future of Regional Cooperation, Coordination and Integration in South Asia.

Connectivity between SAARC Countries

Connectivity has assumed several forms in South Asia. Starting from knowledge connectivity which is made possible through the South Asian University (SAU), SAARC member nations enjoys connectivity through information, communication and technology (ICT), sea (maritime), land and air and above all, energy. By virtue (vice?) of their geographical proximity, it is easier and simpler for the countries of South Asia to connect with each other by developing the necessary infrastructure.

But the half-hearted attempts at completing the ill-thought out infrastructural projects have created several ruckuses in enhancing connectivity in South Asia.

Although there is no denying the fact that India being the largest and prosperous country has effectively help develop roads and other infrastructural projects in all its neighbouring countries but India so far, has not been able to initiate a single project through SAARC which can be said to be truly regional. Almost every project be it Sethua Samudram between India and Sri Lanka or Samjhauta Express between India or Pakistan or hydropower in Bhutan or fisheries in Maldives, is bilateral, not even tri or multilateral.

Another astonishing feature of regional connectivity in South Asia is that it is either India-driven or India-centric. India is central to every project aiming at promoting connectivity in South Asia. It seems that it was Indian influence at work when recently on a question of selling hydropower to Bangladesh, Bhutanese Prime Minister Jigmi Y. Thinley said that Bhutan is already over-capacitated by selling hydropower to India. Similarly, Bangladesh had to keep India in view when Bhutan asked for providing them a transit route through Chittagong and Mongla port. The transit deal which was to be signed in 1980 between Bhutan and Bangladesh could not be signed then because of India's objections. It was signed recently during the Bhutanese Prime Minister Thinley's visit to Bangladesh in January 2011.

Moreover, violent internal disturbances owing to a widespread political instability in South Asian countries and the menace of terrorism combined with severe economic crisis in countries like Pakistan, Afghanistan and Nepal which is totally dependent on external aid and assistance has played a spoilsport in this field. Instead of being inter-dependent, South Asia is totally dependent on external sources for its sustenance. Added to this is the Indo-Pak rivalry which has sidestepped almost every regional initiative taken by the other smaller countries of South Asia.

A very important area which can make a real difference in the living standards of South Asians is energy connectivity between South Asian countries. Imtiaz Alam in 'South Asia Journal' has presented a stark picture of energy situation in South Asia. He estimated that 'as compared to the last two decades, when the energy consumption was 5.8 percent against low energy production of 2.3 percent, the demand for energy is growing at a rate of 9 percent annually, whereas the deficit in energy production has almost doubled in the last decade. Even though South Asia has the lowest per capita consumption of energy in the world (0.45 toe), it is going to have the highest rate of energy consumption by 2010 and beyond. While 60 percent (775.3 million) of its population remains without electricity and is dependent on biomass, the energy intensity (total energy divided by per unit of GDP) is 0.65 toe, as compared to the world average of 0.29 toe—which is much higher' (Alam, 2005).

While India has assisted Bhutan and Nepal to develop their hydropower resources, its inability to forge a viable partnership agreement with them especially Nepal who is reluctant to implement the provisions of the Mahakali treaty paints a grim picture of the future of energy cooperation in South Asia. Similarly, India, Pakistan and Bangladesh which derives almost 73 percent, 69 percent and 92 percent of its energy from coal-based power stations respectively can successfully cooperate in thermal power generation but the political and bureaucratic will is wanting.

Besides, hydro and thermal based energy availability in South Asia, the region can also serve as the corridor for supplying vast amount of gas via pipelines from Persian Gulf and Central Asia. Pakistan can become a viable route for supplying gas to India and beyond. Bangladesh is sitting on substantial gas reserves of approximately 22.9 trillion cubic feet which it can always choose to export to India and Nepal.

There are many practices in a single country of South Asia which have a potential for its effective implementation across South Asia. There is a comprehensive database prepared by Mr. M.W. Leelaratne for a non-governmental organization Practical Action, Sri Lanka in 2009 which has meticulously listed almost all the Renewable Energy practices practised in the countries of South Asia and the potential of their applicability in the entire subcontinent. Mr. Leelaratne has also painstakingly listed the companies who are currently active in providing these renewable energy services to their host countries (Leelaratne, 2009).

A noticeable feat was achieved during the 14th SAARC summit held at New Delhi in 2007 in which it was decided to launch a study for a pilot project connecting one/two hospitals in each of the SAARC countries with 3-4 super-speciality hospitals in India by the end of the year (Jones, 2011). The idea behind this project is to develop an exemplary model for implementing projects at the regional level. It was also decided to replicate its success to expand the scope of regional cooperation to other ICT enabled areas such as education, business process outsourcing and mass communications. In January 2011, SAARC Telemedicine Centre was established in Nepal and it has been decided to link the centre with the Super

Speciality Hospitals in Lucknow and Chandigarh.

SAARC University is another successful example of knowledge connectivity in South Asia and there are many other multilateral projects to connect the neighbours. But the progress is slow and few and far between. The countries in South Asia like to forge linkages with the outside world rather then within the region which is the main cause behind the sluggish progress on a number of infrastructural projects. Without connectivity, it is not possible to develop a feeling of belonging among the general population of South Asia. The lack of interaction will only limit the SAARC and its activities confined to the elite sections of society.

Economic Co-operation in SAARC

South Asian Association for Regional Cooperation (SAARC) was formed in 1985 through an initiative taken by the then Bangladeshi President Ziaur Rahman who envisioned a trade bloc consisting of South Asian countries. Over the years, there have been various economic initiatives taken by the SAARC to enhance economic cooperation among its member countries which can be divided into two phases—pre-liberalization period and post-liberalization period after 1991.

Intra-SAARC trade, as a percentage of South Asia's world trade, increased from 2.42 percent (US$ 1.59 billion) in 1990 to 4.56 percent (US$ 6.53 billion) in 2001 and marginally to 4.7 percent by 2003, mostly attributed to rapid liberalisation under bilateral trade agreements and WTO regimes (Jhamb, 2006). It is to be noted that almost all major regional trading arrangements (RTAs) like South Asian Preferential Trade Area (SAPTA) and South Asian Free Trade Area (SAFTA) were conceived in a post-liberalization era in 1993 and 2004 respectively.

The idea was to subsequently develop South Asian Customs Union, Common Market and Economic Union. South Asia Development Fund (SADF) was created in 1996 after merging of the SAARC Fund for Regional Projects (SFRP) and the SAARC Regional Fund to support industrial development, poverty alleviation, protection of environment, institutional/human resource development and promotion of social and infrastructure development projects in the SAARC region (Narayan, 2010). During the 12th SAARC summit in 2004, SAARC Finance was given the responsibility to examine the concept of a South Asian Development Bank. After the Dhaka summit of 2005, a study for trade in services has been commissioned in 2007.

Before 1991, SAARC member countries followed the autarkic approach towards trade and investments which was inward looking. Moreover, newly-acquired freedom and sovereignty made the South Asian countries more independent in decision making which was earlier tied to the policies of a single colonial power. They have now become free to exercise their choices in choosing their allies and trade partners outside the restrictions of the region. The low levels of development in South Asian countries have further prompted them to look for alternatives outside the region which appears to be more lucrative. Trade promotion activities within the region was also obstructed by the uneven development of economic opportunities in SAARC member countries due to the different economic policies and development models pursued by the different states of South Asia.

After 1991, opportunities for economic integration came in a form of economic reforms and globalization but bilateral differences especially between India and Pakistan and mistrust among the neighbouring countries regarding India's intentions have marred the already bleak prospects of regional economic integration in South Asia. Intra-SAARC trade remained dismally low at a mere 4 percent. However, India experienced a spur in its trade with SAARC bloc as its exports increased from 8 percent to12 percent after 1991 (Jhamb, 2006). Overall, trade in South Asia remained heavily tilted in favour of India which made other neighbouring countries reluctant to transform SAPTA into Free Trade Area (FTA). Despite geographic proximity, weak trade and transit links, infrastructure difficulties, high tariff and non-tariff barriers hamper investments (Jhamb, 2006).

The trade variation is very much evident while comparing the two categories. Also, most of the trade within SAARC region is conducted through a third country. For example, there is a very less trade interaction between India and Pakistan for obvious reasons but Indian goods find their way in Pakistani markets through smuggling via third country and vice versa. Thus, in South Asia, informal trade statistics may be much higher then what are shown in official documents.

Conclusion

The region of South Asia has no regional identity of its own although it does have national identities. Inhabitants of South Asia are the citizens of their countries, then the citizens of the world and lastly, that of the region. South Asians do recognize commonalities with each other in their daily lives, culture and religion but the domain of such complementarities are too narrow to give way for a grand vision of regional integration in the near future. The existing complementarities should not be taken as a necessary step towards regional integration as it may eventually result in loosing even the bare minimum gains in regional cooperation that we have attained so far.

Regional integration should be the ultimate objective of any regional organization and SAARC is no exception but nothing can be expected from the organization whose basic foundations are built on wrong premises ignoring several factors at the same time. In South Asia, three different levels of regional cooperation, coordination and integration are moving simultaneously without any attempts being made to harmonize all the three. Besides, mistrust and misunderstandings between the member nations, bilateral disputes, size disparity, the fear of the so-called Indian hegemony, different levels of socio-economic development are some of the other factors that have obstructed the path of successful regional cooperation in South Asia via SAARC.

The most important thing that should be taken note of is the fact that there is no alternative to SAARC. It can be weak, indolent and slow but it should not be scrapped just because it has failed to achieve its goals. There may be many sceptics raising fingers on the relevance of SAARC, but majority including people and academia wants SAARC to continue. It may not be as fast as EU or ASEAN but its presence so far has not created any hurdles for any member country in the exercise of their sovereignty. There may be several disagreements among its members related to its functioning but SAARC still manage to retain its own separate identity. In future, SAARC will not fade away. It will continue to work gradually towards the ultimate goal of regional integration which is hard to achieve in this region of great diversity. However, common aspirations, concerns and issues will continue to demand a common platform over which it can be discussed and debated. SAARC will thus, continue to have its relevance here.

End Note

1. The areas of cooperation under SAARC are agriculture and rural, biotechnology, culture, economic, energy, environment, funding mechanism, human resource development, information, communication and media, people to people contacts, poverty alleviation, science and technology, security aspects, social development, tourism and trade and finance.

References

Alam, Imtiaz (2005): 'Energy Cooperation in South Asia', *South Asian Journal,* 9.

Guha, Seema (2007): 'SAARC promises more', *Daily News and Analyses,* New Delhi: April 4.

Jhamb, Sujata (2006): 'India's Regional Trading Arrangements', Globalization and Particularism in South Asia', *South Asia Journal,* January-March.

Jhamb, Sujata (2006): "India's Regional Trading Arrangements', Globalization and Particularism in South Asia", *South Asia*

Journal, January-March.

Jones, Kathy (2011): 'SAARC Telemedicine Centre Launched in Nepal', *MedIndia on Mobile,* January 22.

Leelaratne, M.W. (2009): 'Study of Renewable energy Practices in Bangladesh, India, Nepal, Pakistan and Sri Lanka', *Practical Action,* Sri Lanka, January 4.

Narayan, S. (2010): "SAARC and South Asian Economic Integration", in S.D. Muni (ed.), *The Emerging Dimensions of SAARC,* Cambridge University Press India Pvt. Ltd.

Niekerk, Lolette Kritzinger-van, 'Regional Integration: Concepts, Advantages, Disadvantages and Lessons of Experience', http://siteresources.worldbank.org.

Paasi, Ansi and Wolfgang Knapp, (2003): 'A Sense of Place', *ILS NRW*, October 1.

14

Negotiating for Security in South Asia

K.M. Seethi

Concerns for security across the world are now articulated and negotiated within a broad milieu of intra-societal and extra-societal settings. The mainstream scholarship of security studies, during the last two decades, has been grappling with the burgeoning conflicts and tensions in the domestic life of nation-states. However, the emergence, at least since the early 1990s, of a new techno-economic order and its implications for the social architecture of the Third World countries are seldom understood from the point of view of the new security matrix of the post-liberalisation societies where issues of development, livelihood, health, environment etc. are critical in the making of social conflicts and tensions. Admittedly, the state-centric, defence-oriented paradigm of security is seen as inadequate to explain the causative factors of these phenomena as well as to offer solutions to address them (Nye and Jones, 1988; Wolfers, 1962; Clarke 1993; and Buzan, 1991).

However, a new generation of scholars has recently developed concerns for human security, environmental security, gender security, food security, energy security etc. [1] These are also viewed in a broader of context of 'comprehensive security' (Raghavan, 2002; Alagappa, 1998). Though such concerns have ostensible neo-liberal moorings, the domains of engagement (by this scholarship) keep open the possibilities of renegotiation between social security and national security. This is nowhere evident than in the countries in the Third World.

Conceptualising Third World security within the framework offered by the traditional security scholarship is fraught with difficulties. It, therefore, demands a more comprehensive mode in analysing the internal-external dynamics of security in the Third World, particularly in the background of the challenges of globalisation. Security issues nations' deal with today cut across political boundaries, which no state/government can manage single-handedly. This is perceptible in the background of the globalising dynamics of the world economy. It certainly implies that nation-states, particularly of those in the periphery of the global system enjoy less freedom and influence even over those areas that were formerly under their control.

Consequently, the ruling elites of such states are compelled to reassess the role and responsibilities of state, government and civil society in an increasingly neo-liberal mode. The scenario in the Third World unfolds this pattern of development. Security is increasingly under siege within, in virtually all Third World countries, with the collapse of the social security state and the consequent rise of social/ethnic/communal conflicts and tensions. The mounting crisis of social and economic development characterised by uneven growth, poverty, unemployment, resource scarcity, inequitable trade, environmental degradation, health hazards etc. are the consequences of the collapse of the social security state.

Thus, pursuing a more broad level of enquiry would have the virtue of enabling us to see that security in the Third World is much more complex in an increasingly dependent/ interdependent world than the individual states and ruling elites consider it to be. But, for a variety of reasons, the ruling elites in the Third World tend to perpetuate threats and insecurity from internal and regional sources without addressing larger dimensions of them (either global or trans-regional). As such, the complexity of the security problems that all Third World states confront today raises many critical questions about the efficacy of the conventional paradigm in understanding and resolving such issues.

Security is increasingly reckoned today as a positive value mediating tensions and conflicts rather than as a 'negative freedom' merely ensuring 'the absence of the evil of insecurity.' If the notion of security is to be understood to mean a wide range of subjects including human needs, development, social order, identity and adaptability, one should necessarily explore new avenues of maintaining social relations, natural resource base, biodiversity, sustainable environment, health care system etc.

Theorising Regional Security

The post-Cold War situation and the accompanying developments in domestic and international systems, no doubt, pose a major challenge to the conventional paradigm of defence and national security. However, even as critical questions loom large at the global level, some scholars are, still, comfortably engaged in cliché-ridden discourses on regional security. According to them, many of the security problems in the contemporary world are of regional nature and could be addressed as such. This trend could be discernible even earlier.

Way back in the 1980s, Mohammed Ayoob argued that external actors, who might have interest in various regions, would desist from interference; that regional states would have the necessary capability to deal with their own domestic conflicts; and that inter-state stresses and strains are manageable at the regional level itself through sufficient institutional mechanism (Ayoob 1986: 4; Ayoob, 1995). The most forceful expression of such a position could be seen in the Regional Security Complex (RSC) theory of Barry Buzan and Ole Waever (Buzan and Waever, 2003; Buzan, Waever, and Wilde, 1998).

According to them, "the regional level of security has become both more autonomous and more prominent in international politics." They argue that "without superpower rivalry intruding obsessively into all regions, local powers have more room for manoeuvre." To them, the "relative autonomy of regional security constitutes a pattern of international security relations radically different from the rigid structure of superpower bipolarity that defined the Cold War" (Buzan and Waever, 2003).

Advocates of RSC theory claim that this framework would facilitate our understanding of the new international security structure and "the relative balance of power of, and mutual relationship within it between, regionalising and globalising trends". They argue that "since most threats travel more easily over short distances than over long ones, security interdependence is normally patterned into regionally based clusters: security complexes". [2] Buzan and Waever, however, observe: "Security complexes may well be extensively penetrated by the global powers, but their regional dynamics nonetheless have a substantial degree of autonomy from the patterns set by the global powers" (Buzan and Waever, 2003: 4). On the surface level, such arguments may seem to be convincing, but, in reality, they lack any substance, eluding an adequate explanation of the interlocking concerns of global-regional-national paradigms of security.

This paper points to the limitations of conceptualising South Asian security within the framework put across by the regionalists as well as theorists of RSC. It also aims at exploring a more holistic mode in comprehending internal-external dynamics of security in South Asia, particularly in the background of the challenges of globalisation and neo-liberal socio-economic experiences. A major premise of the paper is that security in the era of globalisation is a complex construct unfolding in diverse forms, embedded in internal, bilateral, regional and international settings.

While each setting has its own particular audience, characteristics and concerns, none of them is autonomous by itself (not even relatively autonomous) due to structural compulsions and interlocking/overlapping concerns transcending the fixed paths of national and regional boundaries. This is perhaps most evident in South Asia, which has been in its critical phase of transition during the past two decades. Insofar as security is a much broader category signifying its internal-external dynamics, with different levels of operation within and across national/international systems, problems of threat or insecurity in regions like South Asia need to be addressed within a broad paradigm encompassing issues of development, survival, identity and adaptability. This calls for a new and meaningful negotiation between two contending realms of the state action, i.e., social security vs. national security.

Interlocking Paradigm of Internal-External Security in South Asia

Security issues nations wrestle with today cut across regional/corporeal political boundaries, which no state/government can manage single-handedly or regionally. This is perceptible in the unfolding global scenario, particularly after 9/11 and in the background of the globalising dynamics of world capitalism. [3] The latter certainly implies that nation-states/governments, particularly of those in the periphery of the global system (for instance in South Asia) enjoy less freedom and influence even over those areas/spheres that were formerly under their control.

Consequently, the ruling elites of such states are compelled to reassess the role and responsibilities of state, government and civil society in an increasingly neo-liberal mode. The scenario in South Asia unfolds this pattern of development. The concept of security in the region, just as sovereignty, is challenged by the globalising dynamics of international political economy, open societies and technologies, which make the South Asian states vulnerable to external pressures/forces.

Security is also under siege within, in virtually all South Asian countries, with the collapse of the social security state and the consequent rise of social/ethnic/communal conflicts and movements. The burgeoning crisis of economic development in South Asia characterised by uneven growth, poverty, unemployment, resource scarcity, environmental degradation etc. are the consequences of the collapse of the social security state. [4] Thus, pursuing a more broad level of enquiry would have the virtue of enabling us to see that security in South Asia is much more complex in an increasingly dependent/interdependent world than the individual states and ruling elites consider it to be. But, for a variety of reasons, the ruling elites in South Asia tend to perpetuate threats and insecurity from internal and regional sources only without addressing larger dimensions of them (either global or trans-regional).

The complexity of the security problems that all South Asian states confront in the era of globalisation, raises many critical questions about the efficacy of the regional or bilateral framework alone in understanding and resolving such issues. The narrow, state-centric, militaristic perspective of security is not at all adequate to address the dynamics of the politico-economic processes unfolding in diverse forms across and within national and international systems.

Though the ubiquitous nature of security is generally articulated within the domain of threats/insecurity/danger, [5] it is increasingly reckoned today as a positive value mediating tensions and conflicts rather than as a "negative freedom" merely ensuring "the absence of the evil of insecurity" (Wolfers, 1962: 153). If the notion of security is to be understood to mean a wide range of subjects including human needs, development, social order, identity and adaptability, we should necessarily address the question of maintaining and managing social relations in a way that sustains confidence, cohesion and collectivity, besides providing welfare and social security to the people.

However, the implications of the challenges emanating from the existing global system for South Asia are more profound in that globalisation and its accompanying problems generate (if not exacerbate) tensions within and across nations. It becomes a crucial question when, under neo-liberal conditions, economies of South Asian states have been thoroughly restructured, and the 'welfare state' itself is being rolled back. All South Asian countries have adopted these neo-liberal policies envisaging a minimal role for the state in economic and social life.

As such, the state is more concerned about efficiency, productivity, market and trade than with the questions of welfare, social security and social justice. Neo-liberalism thus seeks deregulation of everything that reduces profit and advocates eliminating the concept of 'public good' pressurising the poorest people in South Asia to seek remedies to their basic problems all by themselves.

It necessarily generates social tensions and social insecurity in South Asia, which, in the ultimate analysis, would ruin the solidarity of political communities. Transitions to neo-liberal economic and political order (obviously under pressure from the international financial institutions) could be more destabilising, and states in South Asia undergoing such transformations are vulnerable to violent/militant/fundamentalist/communal forms of ethno-religious movements (Ahmad and Reifeld, 2004; Phadnis, 1990; Chatterjee, 2001). This is perhaps a critical dimension of globalisation, which has a direct bearing on the security and social development in South Asia.

Insofar as security policies are articulated primarily by the ruling elites and their think-tanks, who have their deep institutional social biases, pressures on them from various internal-external sources are very critical in a situation of increasing trade and techno-economic transactions. The logic of "even flag follows trade" is propagated and applied in global/regional/bilateral ties. In South Asia, India has been the most vociferous advocate of this notion. But, as is evident, even the prospects of a free trade area in South Asia are contingent upon how the ruling elites articulate economic security/insecurity of their particular countries. All South Asian economies are fragile and susceptible to globalising dynamics of world capitalism.

Moreover, regionalism is to be viewed as an adjunct of globalisation not as an antidote to its implications. The ruling elites of all South Asian countries are thus compelled to accommodate the interests and concerns of global capital. But, for small and middle level states in South Asia such as Nepal, Bhutan, Bangladesh and Sri Lanka, India is as much a threat to their survival as it is from other global actors. These interlocking concerns of trade and economic transactions have a bearing on their internal security (in terms of tensions getting accumulated in various sections of each society).

Meanwhile, regional disparities have become a matter of serious concern for each country in South Asia. As such even bilateral free trade arrangements would trigger off conflict/crisis situations in areas where insecurities in the realm of trade and commerce result in serious social imbalances. One instance could be the scenario of plantation sector in Kerala which is concerned about the increasing import of tea, edible oil, rubber etc. from Sri Lanka as part of the implementation of the India-Sri Lanka Free Trade Accord of 1998 (Harilal and Joseph, 2005: 275-89). Though larger economies may ultimately benefit from such trade arrangements (either bilateral or regional), this would have some impact on certain sectors even within such economies.

Energy security is certainly one of the pressing concerns of all countries in the world. The scenario in South Asia is all the more serious given the increasing dependence of the region on other countries for oil and natural gas. But the actual energy requirements of countries like India (Singh, 1997; 342-361) in South Asia are heavily dictated by considerations of global bargain, strategic gains/losses and regional equations. There is very little room for autonomy and manoeuvrability within the region in terms of initiating fresh negotiations with countries like Iran, Turkmenistan, Afghanistan etc. for oil and natural gas.

The proposed gas pipeline projects (for example, Iran-Pakistan-India pipeline, Turkmenistan-Afghanistan-Pakistan pipeline) are susceptible to the pressures being built around India, Iran, Pakistan and Afghanistan by countries such as the United States and Britain. The global non-proliferation regime is also invoked frequently by the Anglo-American group to constrain the manoeuvrability of countries like India in their search for energy security. The whole lot of negotiations before and after the July 18 Indo-US nuclear accord [6] and India's volt-face on the Iranian nuclear question in the International Atomic Energy Agency (IAEA) are indications that the Anglo-American lobby still enjoys enormous influence and power across regional actors.

The so-called "security complexes" are thus more or less patterned into their scheming of global domination and appropriation. In fact, the US willingness to help India develop its nuclear power for energy is primarily to ensure New Delhi's dependence on it for conventional arms, on the one hand, and high-technology for space and nuclear industry, on the other. Washington knows very well that India is so much obsessed with nuclear power (and weapons) and, as a quid pro quo, it expects India obey American insistence on separating civilian and military programmes.

Even as the Indian strategic-nuclear elite is preoccupied with such issues, we seldom address larger security questions emanating from India's fixations on nuclear power (as well as weapons) programme. The social and environmental costs of nuclear power projects are much higher than their actual benefits. Paradoxically, the efforts to ensure energy security are likely to generate new insecurities in the region.

Beyond Region, Beneath National

A critical question while engaging with security in South Asia is whether we can conceptualise 'region' and regional security in a world increasingly dictated by global economic/geo-economic pulls and pressures. Insofar as the political map of all countries in the world has been altering all throughout history—due to various influences from within and outside regions—delineation of regional political map itself could be a problematique: where does one region begin and another end? This can, perhaps, be addressed more clearly in the context of some regions and countries. Turkey's membership in the European Union and Afghanistan's joining of the South Asian grouping are examples of how regions and regional interests are articulated/disarticulated vis-à-vis countries having sensitive geopolitical and geo-economics status.

In South Asia, the geopolitical identities and security perceptions of all countries are intimately bound up with several trans-regional questions encompassing issues of overlapping culture, political economy and security. Migration is one critical question where influx/exodus is prompted by several such factors. The Tibetan refugees in India following the Lhasa uprising, the Afghan refugees in Pakistan following the Soviet intervention and the refugees from Myanmar in Bangladesh are some examples. These phenomena cannot easily be accommodated within a superficial and far-fetched construction of an "Asian supercomplex". [7]

More importantly, all states in South Asia do not share the same perceptions and concerns of regional security. For example, the security concerns of India and Pakistan, the two nuclear weapon powers with a long trajectory of antagonistic relationship, cannot be compared with that of the other non-nuclear weapon states in the region (Ramana and Reddy 2003; Bidwai and Vanaik, 1999). Likewise, membership in a regional entity need not necessarily be viewed uniformly by peoples and ruling elites within the region.

The case of Afghanistan's entry into SAARC is illustrative of this dichotomy in perceptions. Pakistan had, in the 1980s, detested Afghanistan's entry for obvious trans-regional reasons. Why Islamabad welcomed Afghanistan today further explains its trans-regional considerations countersigned by its client Western allies. The entry of China and Japan into SAARC as observers further points to the trans-regional factors in regional bargaining. Thus, it is inconceivable to regard South Asia as "relatively autonomous" (in the manner in which RSC theorists claim), due to ill-defined/ever-changing regional political map, fluctuating membership of regional entities and the direct and indirect involvement of big powers, international financial institutions and other transnational entities.

An important task today is whether we can approach the security problems of South Asia from another angle at this juncture–perhaps going beyond the limits set by the state-centric security/military perspective, on the one hand, and the region-centred notions of conflict-resolution, on the other. This is to be addressed in a broader context of the challenges emanating from globalisation/neo-liberal policies and the consequent disruptive/destabilising tendencies in domestic setting. South Asia can, by no stretch of imagination, be an exception to this process of destabilisation, given the nature of the social and cultural complexities that exist in the region.

Securitisation of Nation-building

The nation-building process in all South Asian countries has been affected by challenges from various religious, ethno-linguistic, tribal and regional identities (Ahmad and Reifeld, 2004; Phadnis, 1990; Chatterjee, 2001). These challenges are not spontaneous in themselves, but caused by both exogenous and endogenous factors having socio-economic and politico-cultural dimensions. The unevenness in the levels of development and the exploitation and suppression (and consequent marginalisation and alienation) of one or a combination of national identities by others generate feelings of insecurity and the underprivileged groups tend to revolt against the dominant identity by initiating protests, struggles and various forms of violence.

In such circumstances of internal challenges, ruling elites

often take refuge in national security. All South Asian countries have such internal challenges with varying degrees of their intensity. A few cases from Pakistan, India and Sri Lanka are illustrative of this scenario.

The state of Pakistan is commonly characterised by the lack of internal legitimacy. This has been perpetuated by a series of crises—systemic and societal—which have had, and continue to have, profound implications for security at the domestic, bilateral, regional and international levels. Domestic challenges to the state's authority have been a frequent cause of insecurity within the country. In fact, the problem of nation-building engulfed the entire Pakistan within a few months of its formation. The emergence of the Punjabi ruling elite with its dominance across all sectors of the system created insecurities among Bengalis (in the erstwhile East Pakistan), Sindhis, Baluchis and Pathans who were underprivileged in the new state (Harrison, 1987: 339-343).

Added to this list are the problems of the Ahmadiya community, Shias, Muhajirs etc. The Pakistani rulers responded to such challenges by invoking a religious argument, emphasising the common bond of Islam. As such Islam was a principle instrument not only for regime legitimacy but for securitising the entire society. This has had obvious implications for Pakistan's foreign and security policies also.

The proclaimed 'unfulfilled religious mission' in Kashmir, the ideological underpinnings of its nuclear programme (Islamic bomb), its consistent support to various brands of Islamic outfits (including Taliban) in Afghanistan, Central Asia, Russia etc. and its public relations campaign through regional entities like Organisation of Islamic Cooperation (OIC) are some of the manifestations of this securitisation project (through the instrument of Islam) of the Pakistani ruling elite. Evidently, this has had trans-regional dimensions and big power involvements. Even today, religion plays an important role in the articulation of security issues in Pakistan, notwithstanding the temporary setbacks since 9/11.

However, the backlash of the US-led campaign and 'war on terror' in the region—undertaken by the military government with very little inhibition–has already surfaced in the form of new and invigorated Islamist/fundamentalist forces in Pakistan. Paradoxically, the ruling elite in Pakistan has been grappling with an unprecedented security situation created by its own deployment of Islam that transgressed the regional boundary of South Asia.

The security scenario in Pakistan is likely to worsen given the magnitude of the legitimacy crisis the military government faces today as well as due to the mounting economic problems within the country. Pakistan's economic and trade liberalisation during the 1990s, initiated largely under the direction and pressure of IMF, has not delivered any good in improving its social and economic development. The conditionalities imposed by the IMF and the World Bank through their Structural Adjustment Programme created a panic situation in the economy and society. Socio-economic indicators have been overturned since the late the 1990s. [8]

The poor economic growth and consequent rise in unemployment together with a relatively high population growth have contributed to a marked increase in the incidence of poverty in Pakistan, particularly since the second half of the 1990s. The prevalence of poverty, which had decreased to 18 percent during the 1980s in Pakistan, has reversed and rose up to 28 percent. The proportion of the population below the poverty line has risen from 20 percent a decade ago to 30 percent today, with the majority of the poor (about 70-80 percent of poor households) living in rural areas.

About two fifths of the population has no access to safe drinking water and more than half has no facility for sanitation. Literacy has remained low and gender disparities in education are glaring. The ruling elite in Pakistan is thus facing a real dilemma as to how to reconcile the mounting internal security problems (generated by its own development trajectory but dictated heavily by global financial institutions) with the professed external security missions abroad (be it in Afghanistan or Kashmir).

From the point of view of security, perhaps the most familiar site of an interlocking conflict situation in South Asia is Kashmir where the two core actors of the region, India and Pakistan, are now not only nuclear weapon states (though they are yet to achieve that status in the perspective and framework of non-proliferation regime), but have their substantial stakes and claims immeasurably associated with domestic political legitimacy. [9] For the ruling elite of Pakistan, Kashmir has been an inevitable symbol of survival (and a raison d'être for perpetual struggle with India) helping to externalise its internal lack of legitimacy.

For this, all successive governments in Pakistan sought the support of extra-regional actors (the United States, Britain, China and a few West Asian states). India, in turn, ensured a countervailing influence with the help of the Soviet Union during the Cold War, and the pattern of building up positions of strength is not substantially different even after the ascendancy of the Anglo-American bloc in the post-Cold War era. Over years, regional balance has been considerably upset, and continues to be dictated, by arms transfers, indigenous arms and missile production and development of Weapons of Mass Destruction. The arms race in South Asia, primarily between India and Pakistan, has assumed a critical dimension in the post-Pokhran-II phase generating new asymmetries and hankering for advanced systems of military power.

The two countries are thus in a competitive race for signing new defence pacts and contracts with the United States, Britain, Russia, France, Israel and China. In spite of some improvement in bilateral ties, both India and Pakistan are obsessed with strategic superiority in the region. Over the last several years, India's military expenditure has skyrocketed in absolute terms. Pakistan too has been diverting the lion's share of its budget for defence without achieving even a modicum of security (leave alone parity) with India (*SIPRI,* 2005). Locked as they are in an unyielding hostility, the ruling elites of India and Pakistan tend to downplay the threats emanating from internal sources such as religious and communal violence, ethnic conflict and from the deep frustration among various sections of the people. Even if they are compelled to respond, the major interventions come in the form of military mobilisation or army action. Kashmir is a glaring case.

It may be noted that a critical phase in the trajectory of the Kashmir question started since the late 1980s when the domestic legitimacy of the government in Jammu and Kashmir (J&K) came under challenge.

This has, ultimately, led to securitising the nation-building project in India. The long spell of manipulation of the J&K politics by regional and national parties, the history of electoral rigging, the persistent neglect of the state's development, deteriorating economic conditions, rising unemployment, worsening industrial backwardness, agricultural stagnation all led to popular discontent, which began to surface for the first time in the late 1980s (Bose, 1997).

The militant organisations which were very few (that too without any mass backing) found this to be an opportunity to make inroads into the society. Pakistan skilfully exploited this volatile situation and began to interpret the question as a renewed case for 'self-determination' and unleashed its campaign. Pakistan's campaign got momentum when India deployed army and paramilitary forces in the state. The encounters between the army and the militants took a heavy toll over the last one and a half decade. There have been innumerable cases of human rights violations by both the militants and the Indian army.

Thus, the human security situation in Kashmir has been worsened with the intervention and involvements of jihadi groups, on the one hand, and Indian and Pakistani states, on the other. The Kashmiri alienation certainly weakened India, but strengthened Pakistan's hands and reinforced jihadi campaign of terror and violence. But this cannot be seen as purely endogenous phenomena. The role of extra-regional powers (such as the US, Britain, China and Saudi Arabia) and non-state actors (jihadi groups and intelligence networks like CIA and ISI) was critical in perpetuating terrorism and fundamentalism across the region.

The Islamic militants in Kashmir had received considerable moral and material support from the US and Britain at a certain stage. More importantly, the fallout of the Afghan resistance movement in Kashmir was long lasting (Seethi, 2002). It is in this background of long years of big power involvement and trans-regional linkages that we must understand the security dynamics of the Kashmir question. The ground situation in Kashmir cannot be ignored altogether even as India and Pakistan are increasingly preoccupied with 'national' security issues. The Kashmir problem is much more complex and profound than what we generally think it as a bilateral or regional issue. It has obvious internal dimensions and perceptible trans-regional implications.

A similar situation exists in the North Eastern region of India also. The simmering discontent in the states of Assam, Manipur and Nagaland has been accentuated by economic downslide, increasing unemployment, rampant corruption, rising rural poverty, fiscal mismanagement and illegal influx from Bangladesh. The insurgency and inter-ethnic conflicts in the region are most often dealt with militarily by India without addressing the fundamental problems of underdevelopment and economic exploitation by various sections, including the local elite (Hazarika, 2002: 28-45). The alienation of land and identity is also at the heart of the problem. Obviously such issues cannot be tackled in a cavalier manner through the military lens.

The ethnic politics and violence in Sri Lanka is another cause of insecurity in South Asia. The volatile situation in the island nation has been created by several critical factors having domestic, bilateral, regional and international dimensions.

From the long trajectory of colonial experience to the postcolonial conditions, the Sri Lankan state and society had undergone different phases of transformation.

Though the underlying causes of Tamil militancy could be explained in the background of the policies of discrimination followed by successive governments in Sri Lanka (such as the Citizenship Act, the Sinhala Only Act, Standardisation in educational institutions, discriminatory recruitment, uneven allocation of development funds, internal colonisation, draconian legislations to deal with militancy etc.), these were accentuated by the economic policy reforms implemented since the late 1970s (Samarasinghe and Coughlan, 1991). And this has obvious internal-external linkages.

Sri Lankan economy was one of the most protectionist and state-controlled in South Asia, at least until the mid-seventies. But, since 1977 Sir Lanka has been under the pressure and dictates of the IMF and the World Bank to introduce a comprehensive market-oriented economic reform programme. [10] While its economy has been restructured and re-oriented, Sri Lankan society and polity experienced a profound and disturbing transformation.

While the new generation of Tamils and Sinhalese began to confront the worst scenario of economic ills, increasing unemployment, decline in agriculture, collapse of traditional industries etc., the government has, from time to time, conveniently bypassed even the constitutional obligations in ensuring social security and social justice. Today, it is a country torn by violent social and ethnic conflicts, rampant corruption, weak and ineffective government, and persistent civil and human rights abuses.

Over years, civil war and military expenditures have mounted draining the vital resources of the country. Sri Lanka is a good example that economic liberalisation can have a very profound and protracted disruptive effect on the structure of society creating an atmosphere of perpetual insecurity (Dunham and Jayasuriya, 1998: 3151-3156; Ravallion and

Jayasuriya, 1988: 247-255). Thus the ethnic question in Sri Lanka cannot merely be understood in terms of the violence perpetrated by the LTTE militants, though serious as it were till the crackdown by the Sri Lankan government.

Besides domestic factors, the role of regional and extra-regional actors such as India (till 1990) and Norway (since the late 1990s), the trans-regional connections of Tamil militants and the increasing involvements and pressures of international financial institutions/trade regimes (both regional and global) in the Sri Lankan economy etc. have had a bearing on the conflict-resolution agenda of the island nation. Thus, the scenario of insecurity in Sri Lanka is to be addressed in an interlocking mode in the domestic, bilateral, regional and international setting.

Countries like Bangladesh, Nepal and Bhutan are also experiencing several types of internal-external challenges. The Maoist insurgency in Nepal with its spill-over across the state boundary has been accentuated by acute economic conditions, rural poverty and mounting unemployment, accompanied by the internal repression unleashed by the Royal army. Though the transition to republicanism and democracy has been facilitated by mass upsurges, resulting in the termination of monarchy, Nepal is still in search of political stability and constitutionalism.

Similarly, Bangladesh has also been rocked by the onslaught of many Islamic militant groups who have taken advantage of the alienation of the people following the lopsided economic liberalisation. The ethnic crisis in Bhutan assumed violent proportions since the 1980s when Ngalong-controlled ruling elite began to adopt polices that amounted to an assimilationist programme. The southern Bhutanese of the Nepali origin rose in revolt, but they had to leave their homeland following state repression (Joseph, 1999). The security/insecurity dilemma of the ruling elite vs. people is a continuing feature of all these countries.

In sum, any solution to the internal-external security

problems of countries in South Asia lies not in charting out the possible course of military strategy, but in developing a mature, comprehensive mode in addressing them, which will deliver good to the people who have been sidelined and taken for a ride with the retreat of the social security state. Unless the ruling elites are able to address such issues, all states in South Asia will have to face more internal security problems emanating from ethno-tribal, communal and other militant/fundamentalist upsurges and, perhaps, the worst scenario of human rights violations and abuses in the region.

Before a quarter century back, the Olof Palme Report had reminded that if all nations continued to pursue security along familiar lines, nothing much could be expected from them in terms of resolving the global issues. It alerted that countries "can only find security in cooperation with their competitors, not against them." Drawing attention to the security dilemma unfolding in the 1980s, the Report stated: "Poverty, unemployment, inflation, the threat of world recession: the problems that make people insecure are economic as well as military." It further noted: "thc contrast between popular expectations for economic growth and improvement in the quality of everyday lives, and the reality of the slow pace of economic development, feeds dissatisfaction, resulting at times in domestic upheaval, and at other times in pressure to divert internal unrest and criticism to external enemies" (Palme Report 1982: 100-124). Similarly, the Brandt Commission had warned: "Deteriorating economic conditions already threaten the political stability of developing countries.

Further decline is likely to cause the disintegration of societies and create conditions of anarchy in many parts of the world." It said: "If each country retreats inwards out of an impulse of self-preservation we shall end up hurting each other and worsening both our collective and individual conditions" (Brandt Report), 1980: 124; The Brandt Commission, 1983: 1). The core message of all these documents is fairly clear: we have to renegotiate between national security and social security. In the era of globalisation, development, identity and interest are dynamically interrelated and are critical in the formulation of security policy in its broad sense (McSweeney, 1999: 197-198, and 208). Thus, security policies should necessarily serve human needs and mediate tensions in an increasingly complex social order. This is particularly relevant to South Asia where security issues can only be addressed in an interlocking paradigm that necessarily cuts across the familiar highways of security complexes.

End Notes

1. The concept of human security finds its clearest expression in a document of the United Nations Development Programme. See UNDP, *Human Development Report 1994* (Oxford: Oxford University Press, 1994); also see Astri Suhrke, "Human Security, and the Interest of States," *Security Dialogue*, 30(3), September 1999.
2. RSC is defined as "a set of units whose major processes of securitisation, desecuritisation, or both are so interlinked that their security problems cannot reasonably be analysed or resolved apart from one another".
3. The impact of globalisation on security has not been subjected to serious studies. Yet a few such attempts are underway. Many argue that globalisation is a critical factor in determining the security agenda of nations, particularly since 1991. See Victor D. Cha, "Globalisation and the Study of International Security," *Journal of Peace Research*, 37(3) 2000, pp. 391-403; Jean-Marie Guehenno, "The Impact of Globalisation on Strategy," *Survival*, 40(4)1998-99, pp. 5-19; Jan Aart Scholte, *Globalisation: A Critical Introduction* (Basingstoke: Macmillan, 2000); and Ian Clark, *Globalisation and International Relations Theory* (Oxford: Oxford University Press, 1999).
4. For details see World Watch Institute, *State of the World: Global Security 2005*(London: Earthscan 2005); World Bank, *World Development Report,2004* (Washington DC: World Bank, 2003); for different perspectives on this theme see Achin Vanaik (ed.), *Globalization and South Asia: Multidimensional Perspectives* (New Delhi: Manohar, 2004).
5. The paradox of security is that "there is never security without

insecurity and that the one always occurs in whatever form with the other." Michael Dillon, *Politics of Security: Towards a Political Philosophy of Continental Thought* (London and New York: Routledge, 1996), p. 18.

6. See "Joint Statement between President George W. Bush and Prime Minister Manmohan Singh," released by the Office of the White House Press Secretary, July 18, 2005; for different perspectives on the theme see Zia Mian, M.V. Ramana, "Feeding the Nuclear Fire," *Economic and Political Weekly*, Vol. XL, No. 35, August 27, 2005, pp. 3808-3811; Amulya K.N. Reddy, "False Assumption of Nuclear Deal", *Economic and Political Weekly*, Vol. XL, No. 35, August 27, 2005, pp. 3812-3815.
7. According to Buzan and Waever, a supercomplex is "a set of RSCs within which the presence of one or more great powers generates relatively high and consistent levels of interregional security dynamics." Buzan and Waever (2003).
8. The unemployment rate has risen from 5.4 percent to 7.8 percent by 2001-2. See Ishrat Hussain, *Pakistan: Economy of An Elitist State* (Karachi: Oxford University Press, 1999); William Easterly, *The Political Economy of Growth Without Development: A Case Study of Pakistan* (Washington DC.: Development Research Group, World Bank, March 2001, *at www.worldbank.org*; also see UNCTAD, *Report on Trade and Development 2002* (New York, 2002), p. 60.
9. For a critical perspective on this question see K.M. Seethi, "Kashmir: Rethinking Security beyond the Line of Control," in Rajen Harshe and K.M. Seethi (eds.), n. 14, pp. 243-259.
10. These global financial institutions today praise Sri Lanka as a positive and instructive example of economic liberalisation. See World Bank, *Sri Lanka: Recapturing Missed Opportunities, World Bank Country Report* (Washington D.C.: World Bank, 2000).

References

Ahmad, Imtiaz and Helmut Reifeld (eds.), (2004): *Lived Islam in South Asia: Adaptation, Accommodation and Conflict*, New Delhi: Social Science Press.

Alagappa, Muthiah (1998): "Asian Practice of Security: Key Features and Explanations," in Muthiah Alagappa (ed.), *Asian Security Practice: Material and Ideational Influences*, Stanford:

Stanford University Press.

Ayoob, Mohammed (ed.), (1986): *Regional Security in the Third World,* London and Sidney: Croom Helm, p. 4.

Ayoob, Mohammed (1995): *Third World Security Predicament: State Making, Regional Conflict, and the International System*, Boulder, CO: Lynne Rienner.

Bidwai, Praful and Achin Vanaik (1999): *South Asia on a Short Fuse: Nuclear Politics and the Future of Global Disarmament,* New Delhi: Oxford University Press.

Bose, Sumatra (1997): *The Challenge in Kashmir: Democracy, Self-Determination and a Just Peace*, New Delhi: Sage Publications.

Buzan, Barry (2003): *People, States and Fear: An Agenda for International Security Studies*, London: Harvester Wheatsheaf,

Buzan, Barry Ole Waever, and Jaap de Wilde (1998): *Security: A New Framework for Analysis* (Boulder, CO: Lynne Rienner.

Buzan, Barry and Ole Waever, (2003): *Regions and Powers: The Structure of International Security,* Cambridge: Cambridge University Press.

Chatterjee, Shibashis (2001): "Ethnic Identity Assertion and Security Issues in South Asia," in Sudhir Jacob George (ed.), *Intra and Inter-State Conflicts in South Asia*, New Delhi: South Asian Publishers, pp. 64-92.

Clarke, Michael (ed.), (1991): *New Perspectives on Security*, London: Brasseys.

Dunham, D. and S. Jayasuriya (1998): "Economic Crisis, Poverty and War in Contemporary Sri Lanka: On Ostriches and Tinderboxes", *Economic and Political Weekly*, December 5, pp. 3151-3156.

Harilal, K.N. and K.J. Joseph, (2005): "Free Trade Area Accord between India and Sri Lanka: Implications for South Indian States," in Rajen Harshe and K.M. Seethi (eds.) *Engaging with the World: Critical Reflections on India's Foreign Policy*, New Delhi: Orient Longman, pp. 275-289.

Harrison, S. (1987): "Ethnicity and the Political Stalemate in Pakistan," in Ali Banuazizi and Myron Weiner (eds.), *The State, Religion and Ethnic Politics: Pakistan, Iran and Afghanistan,* Lahore: Vanguard, pp. 339-343.

Hazarika, Sanjoy (2002): "The Wages of Ad Hocism: Trauma in India's North East," in Delhi Policy Group (ed.), *Comprehensive Security: Perspectives from India's Regions,* New Delhi: DPG, pp. 28-45.

Joseph C. Mathew (1999): *Ethnic Conflict in Bhutan*, New Delhi: Nirala.

McSweeney, Bill (1999): *Security, Identity and Interests: A Sociology of International Relations,* Cambridge: Cambridge University Press, pp. 197-98 and 208.

Nye Jr., Joseph S. and S. Lynn-Jones, (1988): "International Security Studies," *International Security*, 12 (4).

Phadnis, Urmila (1990): *Ethnicity and Nation Building in South Asia,* New Delhi: Sage Publications.

Raghavan, V.R. (ed.), (2002): *Comprehensive Security: Regional Perspectives*, New Delhi: Delhi: Policy Group.

Ramana, M.V. and C. Rammanohar Reddy (eds.), (2003): *Prisoners of the Nuclear Dream,* New Delhi: Orient Longman.

Ravallion, M. and S. Jayasuriya (1988): "Liberalisation and Inequality in Sri Lanka: A Comment", *Journal of Development Economics*, 28: 247-255.

Report of Independent Commission on Disarmament and Security Issues (Palme Report), (1982): *Common Security: A Programme for Disarmament*, London: Pan Books, pp. 100-124.

Report of the Independent Commission on International Development Issues (Brandt Report), (1980): *North-South: A Programme for Survival*, London: Pan Books, p. 124.

Samarasinghe and Coughlan (eds.), (1991): *Economic Dimensions of Ethnic Conflict,* London: Pinter Publishers.

Seethi, K.M. (2002): "The Making of Terrorism and Islamism: Partners in Uneasy Peace and Unholy War," *Indian Journal of Secularism*, 6 (3).

Singh, K.R. (1997): "Meeting Energy Needs: Domestic and Foreign Policy Options," in Foreign Service Institute (ed.), *Indian Foreign Policy: Agenda for the 21st Century*, New Delhi: Konarak, pp. 342-361.

SIPRI Year Book (2005): *Armaments, Disarmament and International Security,* Stockholm: SIPRI.

The Brandt Commission (1983): *Common Crisis: North-South Cooperation for World Recovery*, London: Pan Books, p. 1.

Wolfers, Arnold (1962): *Discord and Collaboration: Essays on International Politics,* Baltimore: Johns Hopkins Press.

15

Regional Integration and Energy Security in South Asia

C. Vinodan and Shibu M. George

The events of recent years have brought South Asia into the international limelight with serious challenges and opportunities for cooperation and development. The renewed interest of major powers to the region, especially the United States and China has added new dimensions to the regional security order. The recent initiatives for the expansion of civil nuclear energy development in South Asia have added new challenges to the emerging regional security architecture. South Asia is home to one of the oldest civilizations of the world. The sub-region includes the countries with "Afghanistan, Bangladesh, Bhutan, India, Maldives, Nepal, Pakistan, and Sri Lanka".

South Asia represents more than one-fifth of humanity, combining population is more than 150 crore growing at the rate of 1.8 percent per annum but has only 4.8 percent of the world's total land area. It is the largest regional grouping in terms of human resources than that of EU and ASEAN. SAARC was formed in 1985 with the goal of raising the living standard of the region's poverty stricken people. But unfortunately it has not been able to leverage its full potential. Geography coupled with high levels of poverty and population density has rendered South Asia especially vulnerable to the impacts of many non-traditional security challenges.

This could compound development problems and further strain the resources needed to sustain growth, urbanization, and industrialization. South Asia is also a conflict prone region subject to continuous political tensions. Tensions have tended to recur periodically and have not allowed an atmosphere of mutual trust to prevail. Nuclearisation and rising influence of non-state actor in the political and security structure of the region further worsened the situation. All these made South Asia a dynamic and complex region that presents the world with significant opportunities and challenges to global security, stability and human well-being.

South Asia is projected to play a key role in the global economy in the twenty-first century. In doing so, the people of this region could address the challenges of poverty, peace, and environmental degradation that confront the world/region today. South Asia is at a historic moment of transforming the socio, economic and political systems and playing a major role not only in the global economy but also in the areas of peace, security and development in the twenty-first century. For the first time in the last 350 years, the international system is undergoing a shift in its centre of gravity from the continents of Europe and North America to the rising Asia.

It is argued that if South Asian countries develop greater regional cooperation and develop an integrated economy, then South Asia could become a major regional force to influence the unfolding global order in the twenty-first century. (Hussain, 2010). This paper argues that greater regional energy cooperation will not only accelerate economic and social development but also ensures peace, stability and security of the region. The paper surveys the current regional energy scenario and arguing that the only way to achieve sustainable energy security is through regional collaboration both within the region as well as with regional neighbours.

The persistent conflict and mutual mistrust among the South Asian States hinders all the potential to develop South Asia as a viable region in the unfolding global scenario. The region has a good resource potential and tremendous scope for energy co-operation, which can play a key role in addressing many of these energy security concerns and in putting it on the path of sustainable development. It is ironic that the record in the area has been so limited and that too in the most basic form of co-operation, i.e. bilateral arrangements between countries.

This paper also puts forth a multi-pronged strategy for sub-regional energy co-operation encompassing softer options aimed at confidence building to more substantial and larger scale co-operation efforts. Delays in decision-making to ensure stronger and mutually beneficial co-operation efforts are associated with high costs not only to the energy sector but also for the entire development agenda of the region.

Energy and Economic Development

Over the past few years, energy security has moved up the global agenda. There are three major reasons for this: first, the impact of high and often volatile energy prices; second, rising Asian powers and the growing demand for energy in India and China; and third, concerns over environmental sustainability and particularly about the global climate change. All these issues are critically important for South Asia in which impressive economic growth has boosted the demand for energy and put corresponding strains on the environment. Energy is a key ingredient of the socio-economic development of any region.

South Asia is not only one of the fastest growing regions in the world; it is also one of the poorest, which thus puts energy at the very heart of the development process in the region. Economic growth and burgeoning populations have put South Asia's energy security in a perilous state. Already energy and power shortages are stunting development in some of the region's least developed locations spurring political insurgences and social dislocation. While the per capita energy consumption is one of the lowest in the world, energy intensity continues to be very high (Ebinger, 2010).

A large portion of the population lacks access to modern sources of energy and depends on traditional sources that are not only inefficient but also have severe health and environmental problems associated with them. Increasing oil import dependency and huge investment needs for energy market development pose a further challenge. Delays in decision making to ensure stronger and mutually beneficial co-operation efforts are associated with high costs not only to the energy sector but also for the entire development agenda of the region.

The availability of energy, that advances—or at least is compatible with—long-term human well being and ecological balance, is a key to the sustainable development of this region. This aspect of sustainable energy supply is critical in the context of South Asian region, as it is poised for higher growth. The growth of economy, along with the population growth in South Asia, has resulted in rapid increase in energy consumption, which is well above the world average of Organization for Economic Cooperation and Development (OECD) rates.

Thus, the South Asian region faces the 'sustainable development' challenge of meeting the rapidly increasing energy demand as well as conserving the natural resources and protecting the environment. With both Indian and Pakistani annual GDP growth rates now in the 6 to 8 percent range and an increasingly young and urbanized population throughout the region, already approaching 1.5 billion people, the South Asian states face instability unless they can quickly find external sources of energy. A future where these states will largely depend on petroleum imports, particularly given the increasingly competitive world market, means that energy security is a primary concern in their foreign policy.

South Asia's mushrooming energy demand is driving growing dependence on energy imports that is likely to accelerate over the next two decades. The result is a deepening sense of energy insecurity that promises to have important long-term geo-political implications for the region. South Asian powers are responding more often with "energy nationalism" than with energy cooperation as they reach out to secure future supplies. The risks of terrorist attacks on vital energy infrastructure are rising, fuelling fears of supply disruptions, and political instability.

South Asia accounts for about 5.9 percent of the world's commercial energy consumption. This does not include the non-commercial energy sources such as wood, animal waste and other biomass that account for more than half of the region's total energy consumption. Despite rapid rise in demand, South Asia continues to average among the lowest per capita energy consumption of 443 kilogram of oil equivalent (kgoe), which amounts to 26.2 percent of the world average of 1,692 kgoe.

However, energy consumption per dollar of GDP is amongst the highest in the world, which indicates the inefficient utilization of energy. Improved energy security is vital to South Asia's economic growth and stability, given that demand for energy in the region is expected to double over the next 15 years. South Asian nations rely heavily on imported energy, yet the region has large amounts of untapped energy resources. The lack of regional energy cooperation and trade compounds the problem. Energy cooperation has the potential to be mutually beneficial for all regional players, with some countries acting as net suppliers, some as net consumers, and some as net distributors of energy. Paucity or lack of capacity, awareness, development of regional energy resources, and sustained regional dialogue is a barrier to increased regional energy cooperation and, therefore, economic growth and could lead to instability.

It is one of the fastest growing regions in the world with an average growth rate of above 6 percent over the past eight years and has demonstrated its potential for emerging as a significant economic entity in the world economic space. (Ahmed, Kelegama and Ghani, 2010). South Asia's prospects changed in the 1980s as it adopted pro-growth policies. It opened up markets to international competition, replaced the public sector with the private sector as the engine of growth, and improved macroeconomic management (Ahmed, 2006). It is now the second-fastest-growing region in the world, after East Asia. Growth rates in South Asia and East Asia appear to be converging. Despite its recent achievements, South Asia continues to be home for more than 40 percent of the world's poor. Large sections of the South Asian people do not have accéss to education, health, nutrition, and such basic amenities like safe drinking water, sanitation and electricity.

Hence the region is facing acute developmental challenges. Two major negative developments have emerged: (a) evidence of growing income inequality in South Asia, and (b) a growing imbalance among regions within countries and among the countries themselves. With fairly large and open borders, the growing imbalances in incomes and opportunities among South Asian countries present similar social and economic problems to the prosperous neighbours as the imbalances within these countries.

South Asia's Current Energy Profile

South Asia is home to 1.5 billion people, about quarter of the world population but has 43 percent of the world's poor and only 2 percent of world's GDP. South Asia, including Afghanistan, Bangladesh, Bhutan, India, Maldives, Nepal, Pakistan, and Sri Lanka, is a region of stark contrasts. One of the most populous regions of the world, it has a relatively small land mass and a high incidence of poverty. Although the South Asian region is a repository of the poorest people in the world, with more people without adequate access to energy [1] than anywhere else in the world, ailed with pressing issues of mortality and health, economically it is also one of the fastest growing regions of the world.

The countries of the region range in size from India, with a population of over 1 billion, to the Maldives with a population of just below 4,00,000. Globalization has brought the region to a crossroads. There are two faces of South Asia. The first South Asia is dynamic, growing rapidly, highly urbanized, and is benefiting from global integration. The second South Asia is largely agricultural, land-locked, exhibits high poverty, suffers from many conflicts, and is lagging (Yoshida, 2008). The divergence between the two faces of South Asia is on the rise. Many policy and institutional constraints contribute to this dichotomy. One important constraint is regional conflict that has made South Asia one of the least integrated regions of the world (Ahmed and Ghani, 2007). While progress has been made in reducing trade barriers with the rest of the world, intraregional trade is a mere 5 percent of total official trade as compared with over 50 percent in East Asia.

While growing populations have kept the per capita income in the countries of South Asia stubbornly below the world average, their economies are expanding and the demand for energy has soared. These nations are trying to meet ambitious poverty reduction goals while their growing industrial, commercial, and transport sectors and urban and middle-income consumers are using energy at unprecedented rates. To unleash the region's economic potential, every country in South Asia will have to face critical decisions, ranging from depleting indigenous energy resources to developing long-term plans to secure national and regional energy needs (Nanda, 2008).

Despite the low rates of economic growth from the fifties to the eighties, the region was undergoing dynamic growth rates averaging 5-6 percent year during the last two decades. The region is also characterized by poverty, illiteracy and low life expectancy. It is the least gender-sensitive region and it has the highest human deprivation, in terms of access to health facilities, safe drinking water, sanitation facilities and opportunity for work and leisure.

At the same time, South Asia is the most militarized region in the developing world, with the highest percentage of GDP going to military expenditures. This implies that a large portion of the income which should have been used for improving the lives of people is being used for armament in view of the contending threat perceptions of the regional countries. South Asia is a dynamic and complex region that presents the world with significant opportunities and challenges to global security, stability and human well-being.

The region is home to India, the world's largest democracy and a rising power with one of fastest growing economies in the world. It is also includes governments struggling to maintain control of their countries, two nuclear powers that coexist in an uneasy and often tense relationship, and conflicts with extremists forces. All of these factors make the region a subject of considerable strategic importance to the major powers and the larger international community.

The recent trends in economic growth shows the demand for energy in South Asia is going to increase steadily and substantially for the foreseeable future. According to the International Energy Agency forecast for the region, this demand will probably more than double over the next 30 years—Indian requirements alone are projected to increase more than 140 percent—compared with an increase of only 55 percent for the world as a whole (IEA, 2009). This rapid rise in energy needs is, of course, a necessary corollary of continued economic growth. It is well known that although countries may differ in their ratios of energy consumption to economic output, or *energy intensity*, there is a clear and unavoidable correlation between the two (Singh and Pannu, 2009).

Moreover, this energy intensity tends to increase in any given economy as the focus of production shifts from the agricultural to the manufacturing and service sectors (Newberry, 2007)—exactly what economists say has to happen in countries such as Pakistan and India if recent growth rates are to be sustained (Ahmed and Ghani, 2007). Even without such a shift, two other factors suggest that energy intensity in South Asia will have to increase nevertheless. One is that, according to the World Bank, unreliable supplies of energy, particularly electricity, are already the most important factor in the high cost of doing business in South Asia compared to other regions. Reducing such costs is one of the essentials of sustained growth (ibid).

The other is that whether or not economic activity shifts away from the agricultural sector, the pressing need to improve standards of living in the backward regions will require a range of specific measures such as rural electrification, improving transportation of crops to markets, better medical care, clean drinking water, and access to cleaner cooking and heating fuels, all of which will further drive up energy requirements.

Another major reason for the energy crisis in South Asia is the volatile situation in the energy pricing especially in the international crude oil prices. Energy economists say that we are never going to see the days of cheap energy again (Moran and Russel, 2008). Strategists in the developed countries have generally accepted such an upward trend and started diversifying their energy security policy. The reality is that no South Asian country is going to be able to meet its energy needs entirely from within its own domestic resources.

Most of the countries in the region have energy endowments of one kind or another: India has significant coal deposits, Bangladesh has natural gas, and the countries across which the Himalayas and Hindu Kush lie have substantial undeveloped hydroelectric potential. But development of the transportation sector in these countries is going to depend on oil, which no South Asian country has yet discovered in any substantial quantities. In short, none of the South Asian countries possesses the entire range of energy resources needed to meet its development objectives.

The obvious conclusion is that energy trade must be part of the solution to South Asia's continued development. A large share of that trade must clearly come from outside the region. After all, the South Asian countries collectively are now able to meet less than 30 percent of their combined oil needs from within the region; if increasing demand is to be met at all, most of the new supplies will have to come from the Persian Gulf or beyond (Ebinger, 2010).

However, in addition to external sources of supply,

international energy experts increasingly believe that greater cooperation within South Asia and with adjoining regions would be one of the most effective ways—perhaps an indispensable way—to deal with the regional energy deficit (Newberry, 2007). This emphasis on regional cooperation has caught the attention not only of academic economists but of the official development community as well.

The rationale for looking to regional cooperation as the key to meeting South Asia's burgeoning energy needs is laid out in a recent World Bank study that describes such an approach as providing a "logical and rational public policy choice a win-win situation to all the participants" (Krishnaswamy, 2007). The study points out that the fundamental economic conditions for trade exist: some countries in and near South Asia (specifically Bhutan, Burma, Iran, Kyrgyzstan, Nepal, Tajikistan, and Turkmenistan) have more fossil or hydropower resources than they can use, while others (Afghanistan, Bangladesh, India, Pakistan, and Sri Lanka) face a widening gap between demand and supply.

In 2000, energy cooperation was brought directly under the purview of the SAARC Integrated Programme of Action. Under the programme of Action the subject was being dealt with by the Technical Committee on Science and Technology. In view of the increasing importance of cooperation in this sector SAARC established a Technical Committee on Energy in 2000.

Recognizing that this vital area needed intensive attention, the 12th SAARC summit held in Islamabad in 2004, set up a specialized Working Group on Energy to undertake a focused study of various possibilities for cooperation in the field of energy including the concept of an 'energy ring' that would include transnational energy lines for trade in electricity, gas and oil. The subsequent induction of Afghanistan in SAARC introduced an added dimension to the issue as that country, connecting Central Asia with South Asia, can play an important role in that sector. Special attention was given to the development of environment friendly energy, to promote non conventional and renewable sources of energy for economic development and poverty reduction, and to place special focus on energy availability in rural areas.

It has formulated the terms of reference for study on energy trade in the region (Sami 2008). At the SAARC Energy Minister's conference in Islamabad in September-October 2005, it was agreed to establish a SAARC Energy Centre (SENTER) in Islamabad, Pakistan. The primary objective for the establishment of SENTER is to have a regional institution of excellence for the initiation, coordination and facilitation of SAARC programmes in Energy. The promotion of private sector investment and participation in energy activities of the region assumed top most priority.

This is to convert the saving potentials into investments in energy projects (electricity, liquefied natural gas) through both the public as well as the private sectors, as integrated favourable investment policy should be formulated that can attract region's private investors to invest in energy projects. The SAARC countries current dependence on outside sources for their growing energy requirements can be removed by better management efficiency and stronger cooperation amongst themselves (Nitya, 2008). The 16th SAARC Summit also focused on issues of energy security and regional energy cooperation, under the sprightly banner of 'A Green and Happy South Asia. [2]

Energy Security: Growing Needs of South Asia

Energy security—the ability of a nation to secure sustainable energy supplies to meet its energy needs at reasonable prices—has become a major energy policy imperative. The energy security situation is especially acute for South Asian nations, where energy sectors must typically grow by 2-3 percent over the GDP growth rate simply to sustain the economy. Each country in the region must therefore address critical energy concerns that will determine not only its national economic expansion plans, but also its ability to bring underserved local populations onto the grid and into the economy.

Energy requirements in South Asia are growing in industry, transport, residential, agriculture and other areas (Siddiqi, 2007). Higher economic growth and demographic pressures have resulted in phenomenal consumption of commercial energy in South Asia. Although significant progress has been made during the last five decades in increasing energy supply, lack of access to low cost, clean, and reliable sources of energy still remains the daunting challenge to sustain future growth momentum in South Asia.

The region is utterly vulnerable owing to inadequate regional energy endowments and overwhelming dependence on imported fossil fuels. Heavy dependence on traditional fuels, such as agricultural and animal residues, and on fossil fuels like coal and oil in the region also points to implicit environmental and health costs. Importantly, from a broader perspective of deregulated global energy scenario, South Asian energy systems are relatively inefficient. Such outmoded energy supply chain infrastructures and inward looking policies are incapable of absorbing market volatility and also protecting the highly vulnerable section of society from the cataclysmic effects of energy deregulation.

Considering the critical role of energy in modern economies and the pernicious effects of any disruption thereby, it is imperative to ensure energy security for the well being of the people and the economy. In South Asia, the challenge is not only due to the limited reserves of energy resources within the sovereign jurisdiction of nations, but also due to the absence of energy trade and investment in the region which have often resulted in sub-optimal and high cost decision making at the national level. In the present era of regionalism, it has been observed that the rationalisation and harmonisation of regional economies across the world have started with cooperative ventures involving specific commodity sectors.

Table 15.1: Selected Energy Indicators of South Asia, 2006

Indicators/Country	Bangladesh	India	Nepal
Population (million)	153.28	1134.40	27.13
GDP (billion 2000 US$)	61.36	644.10	6.35
GDP ppp (billion 2000 US$)	259.08	3362.05	37.43
Energy Production (Mtoe)	19.31	419.04	8.15
Net Imports (Mtoe)	4.58	121.60	1.02
Total Primary Energy Supply (TPES) (Mtoe)	24.19	537.31	9.17
Electricity Consumption[a] (TWh)	20.78	525.93	1.88
CO_2 Emissions[b] (Mt of CO_2)	36.34	1147.46	3.05
TPES /Pop (toe/capita)	0.17	0.49	0.34
TPES/GDP (toe/000 2000 US$)	0.39	0.83	1.44
TPES/GDP (ppp) (toe/000 2000 US$ ppp)	0.09	0.16	0.24
Electricity Consumption/Pop (Kwh/capita)	147.00	480.00	69.00
CO_2/TPES (t CO_2/toe)	1.50	2.14	0.33
CO_2/Populations (t CO_2/capita)	0.26	1.05	0.11
CO_2/GDP (kg CO_2/2000 US$)	0.59	1.78	0.48
CO_2/GDP(kg CO_2/2000US$ ppp)	0.14	0.34	0.08

Contd...

Indicators/Country	Pakistan	Sri Lanka
Population (million)	158.08	19.12
GDP (billion 2000 US$)	92.77	19.66
GDP ppp (billion 2000 US$)	328.49	80.22
Energy Production (Mtoe)	61.27	5.25
Net Imports (Mtoe)	15.54	4.32
Total Primary Energy Supply (TPES) (Mtoe)	76.33	9.37
Electricity Consumption[a] (TWh)	71.07	7.43
CO_2 Emissions[b] (Mt of CO_2)	118.40	12.27
TPES /Pop (toe/capita)	0.49	0.48
TPES/GDP (toe/000 2000 US$)	0.82	0.48
TPES/GDP (ppp) (toe/000 2000 US$ ppp)	0.23	0.12

Contd...

Electricity Consumption/Pop (Kwh/capita)	456.00	379.00
CO_2/TPES (t CO_2/toe)	1.55	1.31
CO_2/Populations (t CO_2/capita)	0.76	0.63
CO_2/GDP (kg CO_2/2000 US$)	1.28	0.62
CO_2/GDP(kg CO_2/2000US$ ppp)	0.36	0.15

Notes: (a) Gross Production + imports - exports - transmission/distribution losses. (b) CO_2 emissions from fuel combustion only. Emissions are calculated using IEA's energy balance and the revised 1996 IPCC guidelines.
Source: RIS based on International Energy Agency (IEA), (2007).

In this respect, given complementary energy resource endowments, unlocking the vast untapped energy trading potentials through regionally integrated energy cooperation assumes also work significance, as it would not only enhance the region's perennial energy security but also work as a hook for broad based regional economic integration.

Between 1980 and 1990, energy use [3] or primary energy consumption in South Asia increased by 51 percent and later went up by 53 percent, at an annual average growth rate of 4.1 percent, between 1990 and 2007. Energy use is closely linked with economic growth and other vital developmental parameters. South Asia continues to average among the lowest level of per capita energy consumption in the world, despite the fact that it has the highest levels of energy consumption per unit of GDP. [4] Electricity consumption, by far the most convenient form of energy, has grown significantly over the years. Yet, the per capita electricity consumption in South Asia remains low in comparison with other region. Electricity is a high quality form of energy and many of the services it provides, such as the use of computers, televisions and instant communications, cannot be provided by any other form of energy. The demand for electricity far exceeds available capacity in many parts of South Asia and is expected to more than double by 2020.

One of the advantages of the electricity is that essentially any source of energy can be used to generate it. The largest source of electricity is coal, whereas natural gas is the fuel of choice in Bangladesh and Pakistan. Nepal and Bhutan depend mainly on hydro power, whereas Afghanistan, the Maldives and Sri Lanka are primarily using oil products for electricity generation. Nepal and Bhutan have great potential for generating hydro power far beyond their foreseeable domestic demand. Nepal is estimated to have an annual potential above 43,000 MW [5] that can be economically developed, Bhutan has an estimated potential of about 16,000MW. [6]

To put these numbers in some perspective, the total generation capacity in Bangladesh was 4,710 MW in 2005. Only about one percent of potential hydro power potential of Nepal and Bhutan has been developed so far. The two countries are in an excellent position to contribute to the energy security of the entire region by developing their hydro power resources for domestic use, as well as for export to India and possibly to Bangladesh and Pakistan. Until 1998, Bhutan was importing electricity from India. Following the completion with India's help of the Chukha Power Plant, Bhutan in turn now exports electricity to India.

This export has become a single largest source of revenue for Bhutan. The International Finance Cooperation is providing financial assistance to Nepal to build its second private hydro power plant. The electricity is to be used within Nepal. Negotiations between India and Nepal for the sale of electricity from the latter have been going on for a decade, but the two countries have not been able to agree on a mutually beneficial framework.

Energy shortfalls typically in the form of widespread outages of electricity are one of the major problems in South Asia. Electricity transmission and distribution losses as percentage of output are highest in the South Asian region. [7] Also, there is huge economic loss due to poor power quality in some South Asian countries, namely in Nepal and Bangladesh. It is estimated that due to poor power quality, industrial sector in Nepal lost US$ 24 million consisting of 4.43 percent of the industrial GDP or 0.74 percent of the national GDP in the year 2001. Similarly, for Bangladesh the estimated loss suffered by industry is Tk.47 billion of production loss, Tk.12 billion of value added loss, Tk.27 billion of foreign exchange loss, and reduced GDP growth by about half a percentage point (Raina 2005). In addition, since vast majority of population, especially in rural areas of the region do not have access to electricity, providing access at affordable cost is a daunting task for all countries of the region.

South Asia's energy requirements are growing in Industry, transport, residential, agriculture, and other areas. Each of the sectors uses a different mix of energy sources and is growing at different rates. The growing industries of the region need more electricity, oil products, and natural gas. The transportation sector will continue to require larger accounts of oil products for at least the next two decades during which time alternative fuels such as natural gas can begin to make inroads. The demand for electricity in the residential sector will continue to grow, along with the demand for cooking and heating fuels in the northern areas of South Asia. In addition, the rural areas now require access to electricity and also to cleaner replacement for the firewood and other biomass fuels that have traditionally been used. These demand can best be met by solar and wind energy where available and by natural gas or liquefied petroleum gas (LPG).

Energy Mix and Consumption Pattern in South Asia

Commercial sources dominate South Asia's energy consumption pattern. In 2007, South Asia's total energy consumption was 584.6 mtoe, of which the consumption of commercial energy was 420.2 mtoe, indicating a share of nearly 72 percent. The remaining share 28 percent is accounted by traditional fuels like biomass. Coal dominates South Asia's commercial energy mix by virtue of it having the major share in the energy mix of the largest consuming country of the region (India). The respective shares of coal, oil products, natural gas, hydroelectricity, nuclear, and biomass in the region's energy mix was 41 percent, 35 percent, 15 percent, 6.1 percent, 1.1 percent and 1.8 percent, respectively, in the year 2007.

Importantly, there are significant variations in the energy consumption pattern of individual countries in the region. Bangladesh energy mix, for example, is dominated by natural gas (68 percent in 2007), while India relies heavily on coal (50 percent in 2007). Afghanistan, Maldives, Nepal, and Sri Lanka are overwhelmingly dependent on oil products (27 percent), natural gas (55 percent), and primary electricity (12 percent). The Himalayan kingdom of Bhutan has the highest share of hydroelectricity power (70 percent) in its energy consumption mix.

Petroleum consumption in industrial and transportation sectors is increasing South Asia's dependence on imported oil and natural gas. The growing demand for fuels in the industrial and transport sectors in view of industrialization and newer development projects has increased the demand for commercial fuels significantly in South Asia in recent times. The transportation sector primarily depends on oil products for fuel. The rise of middle income groups and the easy accessibility of car finance in most of the countries in the region have put forth strong pressure for personal transportation and consequent pressure on fuel demand.

Although the demand for oil is higher now, South Asia's oil reserves are limited (Chaudhury, 2009: 139). Import of petroleum products to meet the growing demand of energy is going to be on the increase which will put additional pressure on the balance of payment. Such a state of over dependency is likely to affect adversely the energy security of the countries in the region. South Asia is a net importer of oil. In recent years, oil imports have increased due to increasing consumption and declining regional production.

Over the period 1990-2007, South Asian oil consumption,

led by India, grew by about 111 percent. In 2007, oil consumption in South Asia around 4.3 million barrels per day (mb/d) against regional production of merely 0.98 mb/d making the region a net importer of around 3.3 mb/d. The major factor behind the growth of oil consumption is the growing demand for transportation fuel and fuel for generating electricity for industrial purpose. Consumption of natural gas in South Asia has increased considerably in recent years by almost 195 percent between 1990 and 2007. In 2007, South Asia produced 7.8 bcf (billion cubic feet) of natural gas and consumed 8.7 bcf. The largest regional producer and consumer is India, which also has started importing liquefied natural gas (LNG) from Qatar since 2002.

In 2006, the total installed power generation capacity in South Asia was about 728 billion kilowatt hours (bKwh), dominated by India (85 percent share). With 12 percent share in total installed power generation capacity in South Asia, Pakistan comes next. Rest of the South Asian countries have a negligible share. Conventional thermal power plants contribute almost 81 percent of the total generation capacity of the region, whereas the rest is shared by hydroelectric plants (16 percent), nuclear plants (2 percent), and renewable-based plants (1 percent).

Major component of power generation in South Asia is thermal power plants, mainly because of heavy dependence on coal-based generation by India. Bangladesh and Pakistan also have major generation coming from thermal power plants. In case of Nepal, Bhutan and Sri Lanka, the generation-mix is dominated by hydropower. Out of the above installed power generation capacity renewable energy constitutes only about 1 percent of the installed capacity today. It can also be added that only India and Pakistan have power generation from diverse sources.

South Asia is also a region with the highest growth rates in the consumption of commercial energy. Biomass in the form of fuel wood, agricultural waste, and animal waste remains as a primary fuel for the majority of poor people in region, although the spread of electricity and the penetration of fuel products for lighting and cooking, has led to a gradual reduction in the share of biomass in most of these countries. Much of the imported oil products are used as fuel for the transportation sector as well as to meet some urban cooking needs (Regional Report, 2009).

Prospects of Energy Trade and Cooperation in South Asia

There are substantial gains from the potential regional energy trade in South Asia. All countries of the South Asian region have an inadequate supply, current or potential, of some energy form or other, while some of the countries have certain forms of energy far in excess of their demand in the foreseeable future. The countries are eager to have common or collaborative plans to meet the shortages and also to deal with the surplus energy. The energy security plans of these countries indicate that all the countries are eager to develop and increase the energy trade within the region.

The policy makers and strategic planners in South Asia are well acknowledged the vast energy potential of the South Asian region. Afghanistan and Nepal are sitting on water resources that could potentially generate some 24,000 megawatts of electricity from Afghanistan and an estimated 83,000 megawatts from Nepal. These countries together account for 40 percent of South Asia's presently installed capacity. Bangladesh, India, and Pakistan are all power-deficit countries, especially India. The growing electricity constraint is threatening the ability to sustain rapid growth.

Yet, less than 1 percent of the region's electricity-generating potential has been used so far. The reason for this energy constraint is the lack of cooperation and absence of energy trade among South Asian countries. Indeed, if one were to imagine South Asia without borders, perhaps the highest priority investment would have gone to develop the hydropower resources. While all countries would benefit from the development of South Asia's hydropower resources, Afghanistan and Nepal, the two poorest South Asian countries, would benefit most.

After decades of insignificant cross-country electricity trade and the absence of any trade in natural gas through pipelines, regional political leaders and businesspeople recently have evinced a great deal of interest and enthusiasm in cross-border electricity and gas trade, not only within South Asia but also with its neighbours in the West (Central Asia and Iran) and in the East (Myanmar) (Nanda, 2008). South Asia has two regional energy clusters. The eastern market includes India, Bangladesh, Bhutan, Nepal, and Sri Lanka, extending to Myanmar; the western market includes Afghanistan, India, and Pakistan, extending to Central Asia and Iran.

India bridges these two clusters. Some activities are under way, including a successful hydropower trade between Bhutan and India in the eastern market and an ongoing project in the western market that will bring electricity from Tajikistan and Kyrgyzstan to Afghanistan and Pakistan. [8] The current regional security environment is not favourable to exploit the full potential of the region. It will mainly rely on the will of the policymaker and administrators to take concrete and decisive actions in the energy front.

It is argued that the governments of the region need to continue reducing political and security tensions; consider energy trade as an enhancement of energy security and political and economic cooperation; continue energy sector reforms; improve commercial performance of the utilities; improve the credibility, competence, and accountability of regulation; adopt sustainable (cost-reflective) tariffs and a social protection framework; promote commercial approach to energy trade; encourage private sector participation in the form of public-private partnership (PPP) structures in cross-border investments; help the transit countries (especially Afghanistan) integrate; reach water-sharing agreements; seek accession to international agreements (such as the Energy Charter Treaty);

strengthen regional institutions at both political and technical levels; and identify priority trade-oriented investment projects and pursue their implementation.

The success of the India-Bhutan electricity trade should offer useful lessons to other countries in the region. Restrictions in transport border crossings are a major constraint to global and intraregional trade in South Asia. Removing these restrictions would boost trade within South Asia as well as lower costs for international trade in general, as many landlocked countries and regions would benefit from access to the closest ports. Currently, efforts to improve trade facilitation and transport networks are being pursued in a fragmented manner, and where cross-border issues are involved, little cooperation exists. Establishing corridor-based approaches for improving the trade, such as a transport arrangement for intraregional trade, would be essential to improve the efficiency of regional transport and to reduce trade costs.

Some of the larger countries are also proposing joint action to explore and procure energy resources, especially hydrocarbon resources, from areas outside the region. There is also the possibility of jointly procuring some fuels, which could result in economies of scale in transaction and transportation costs (Chaudhury, 2009). In this respect, given complementary energy resource endowments, unlocking the vast untapped energy trading potentials through regionally integrated energy cooperation assumes significance, as it would not only enhance the regions perennial energy security but also work as a hook for broad based regional economic integration (RIS 115, 2008).

Regional trade in energy is practically non-existent in South Asia, with the exception of small pockets of bilateral power exchanges between Nepal, Bhutan, and India. Longstanding disputes, political exigencies, and mistrust between the countries of the region have effectively blocked even modest efforts to encourage regional energy trade (SARI,

2009). In view of the ever increasing global demand for energy and steep rise in price, the South Asian countries are finding it difficult to adequately meet the demand of an expanding economy on one hand and the rising aspirations of its people on the other. To enhance energy security and to achieve sustained economic growth in South Asia it will be essential to maintain energy supply in an efficient and reliable manner. This will call for the long term vision to develop an integrated energy market in the region (Sami, 2008: 150).

With the geography of energy supply options not confined to political boundaries, South Asian countries would need to depend upon each other for the cheapest and cleanest energy sources. Cross-border energy supply would not only provide diversified energy supply, which is a key component of energy security, but also foster regional cooperation. Therefore, the importance of regional integrated energy cooperation is vital to South Asia's development and prosperity.

A common regional energy market would provide, among others: (i) tremendous scope for regional economic development through increasing access to reliable and affordable energy sources, (ii) enormous financial gains through freer energy trade, (iii) mainstreaming energy trade as a fundamental aspect of regional resource management to foster broader regional integration, (iv) more attractive markets for private investment, (v) better use of global investment flows and technology transfers, (vi) easier integration with global markets, and (vii) reduced cost of supply for all participants. The South Asia countries have huge potential for energy trade by developing integrated energy infrastructures such as power grids and pipelines.

The general consensus is that deeper regional energy cooperation would mitigate the localized shortfalls in energy production through regional transfers of bulk energy from areas of surplus or potential surplus to the deficit regions in South Asia (RIS 122, 2008). The two generally accepted power trading mechanisms are: bilateral power trade and pool based power trade. Cross border power trade on a bilateral basis already takes place widely between India and Bhutan and to a certain extent between India and Nepal.

Currently the power trading is in its infancy in South Asia region. Whatever trade takes place as of today are basically bilateral exchanges or apportioning of power from surplus areas to temporarily needy regions. In most cases, the tariff has been fixed mostly on the basis of negotiations and to a large extent determined by political consideration, diplomatic goodwill and convenience. Such an ad hoc arrangement based on negotiations and goodwill could work in the past mainly because the quantum of power purchase was limited. However, in years to come the size of power purchase would be substantial.

Power trading will be on bulk supply and have to have a more detailed framework of contracts and operating procedures. The proposed Iran-Pakistan-India pipeline, Turkmenistan-Afghanistan-Pakistan-India pipeline, Oman-India sub-sea pipeline, and the Myanmar-Bangladesh-India pipeline projects offer promises for energy security; however, each has been plagued with political, economic, or security obstacles that have prevented tangible progress.

In South Asia, stark institutional barriers hinder regional cooperation in energy. Besides the SAARC, there is no institutional and legal framework for the promotion of regional energy trade. Though SAARC has one of its objectives of strengthening and expanding electricity grid in the region, no progress has been made so far in inter-regional power trade. At the moment there is no institution outside of SAARC framework to play the coordination and energy trade enhancement role. Although there is huge need and possibility regional trade in energy, the existing regulatory and pricing policies do not address the eventuality of regional trade.

Regulatory barriers and distorted energy pricing and subsidy regimes in countries of the region precludes trading of energy at the commercial terms, as these entities, which are selling energy at subsidized rates, will have to bear the energy at cost, with the consequence of being impacted negatively on the financial health of the purchasing entity.

Conclusions

Energy security has moved in to the forefront of the development discourse in South Asia due to the impressive growth performance as well as the continued socio-economic backwardness of the region. In South Asia, poverty alleviation is an issue of high economic development priority. The supply of reliable, sustainable energy, particularly electricity to the rural poor households is critical to addressing this issue. With less than 25 percent of the population having access to electricity in rural South Asia, the task ahead is gigantic. Rising economies and burgeoning populations have put South Asia's energy security in a perilous state.

Already energy and power shortages are stunting development in some of the region's least developed locations spurring political insurgences and social dislocation. South Asia is projected to play a major role in global energy markets over the next several decades, with India alone expected to become the world's third largest importer of petroleum by 2030. Satisfying the region's growing demands will require a heightened degree of energy interdependence among historically antagonistic states.

Consequently, like it or not, regional leaders will face a trade-off between traditional desires for energy self-sufficiency and the ambitious development targets that they have set for themselves. Achieving such growth, therefore, requires that India, Pakistan, and the other countries of South Asia first address the persistent international disputes that hamper cross-border energy trade, establish effective control over presently ungoverned areas, reorient the missions of military forces to some extent, and develop a better understanding of the effects that energy interdependence will have on broader relations with neighbours.

With the precarious energy situation in the region and unprecedented increases in international oil prices seen in recent times, it is high time for policy makers, financing institutions, NGOs and the civil society to push for a greater integration of the energy sector. To pursue energy security, the countries of the region needs to ensure that energy supplies are available, sufficient, affordable and sustainable. This will mean taking a broad range of measures: conserving and raising energy efficiency; rationalizing pricing and taxation systems; improving energy sector governance; and diversifying energy supplies, in particular making greater use of alternative and renewable resources.

The huge rural population and the economics of South Asia constitute two main challenges facing access to electricity in rural areas. Successful models of rural energy practices are available, which can be considered for replication in other countries of the region. Regional cooperation can be an ideal framework for such cooperation. Greater cooperation within South Asia and with adjoining regions would be one of the most effective ways to deal with the regional energy deficit.

End Notes

1. The term energy 'access' generally means access to clean, affordable and Reliable energy services, covering both urban and rural areas, and all forms of energy including traditional sources (Ailawadi and Bhattacharyya 2006). See for more details Ailawadi, V.S. and S.C. Bhattacharyya (2006): "Access to energy services by the poor in India: Current situation and need for alternative strategies", *Natural Resources Forum*, 30, pp. 2-14.
2. Declarations of the Sixteenth SAARC Summit, Thimbu, 28-29 APRIL 2010.
3. Energy use refers to use of primary energy before transformation to other end-use fuels, which is equal to indigenous production plus imports and stock changes, minus exports and fuels supplied to ships and aircraft engaged in international transport.
4. GDP per unit of energy use is the ppp GDP per kilogram of oil equivalent of energy use. GDP (ppp) is gross domestic product

converted to 2000 constant international dollars using purchasing power parity rates. An international dollar has the same purchasing power over GDP as a US dollar has in the United State.

5. South Asia Regional Initiative/ Energy (SARIE/E), "Energy Security for South Asia", New Delhi: US Agency for International Development, 2005.
6. "Hydroelectric power Project of Bhutan", Bhutan News Online, www.bhutannewsonline.com/hydro_electricity; html accessed on September 26, 2006.
7. Electric power transmission and distribution losses include losses in transmission between sources of supply and points of distribution and in the distribution to consumers including pilferage.
8. This is the Central Asia-South Asia (CASA) energy project that seeks to sell 1,000 megawatts of surplus power from Tajikistan and Kyrgyzstan to Afghanistan and Pakistan. The project is being developed in cooperation with a number of multilateral financial institutions, including the World Bank.

References

Ebinger, Charles K. (2010): *Energy and Security in South Asia: Cooperation or Conflict?,* Brookings Institution Press, Washington.

Ailawadi, V.S. and S.C. Bhattacharyya (2006): "Access to energy services by the poor in India: Current situation and need for alternative strategies", *Natural Resources Forum*, 30, pp. 2-14.

Newberry, David (2007): "Power Sector Reform, Private Investment, and Regional Cooperation," in *South Asia: Growth and Regional Integration*, ed. Sadiq Ahmed and Ejez Ghani, Washington, DC: International Bank for Reconstruction and Development/World Bank.

Moran, Daniel and Russel, James A. (2008): "The Militarization of Energy Security," *Strategic Insights* 7, No. 1 (February, available at,
<www.ccc.nps.navy.mil/si/2008/ Feb/moranFeb08.pdf>.

Krishnaswamy, Venkataraman (2007): *Potential and Prospects for Regional Energy Trade in the South Asia Region,* Washington, DC: World Bank.

Sami, C. and M. Shafi, (2008): "Connectivity through Energy and

Harmonization of Resources", (eds.), in Deepankar Banerjee and N. Manoharan, *SAARC: Towards Greater Connectivity,* New Delhi: KAS Publication Series.

Nanda, Nitya (2008): "Energy Cooperation and Connectivity in South Asia", (eds.), in Deepankar Banerjee and N. Manoharan, *SAARC: Towards Greater Connectivity,* New Delhi: KAS Publication Series.

Siddiqi, Toufiq A. (2007): "Viable and Environment-Friendly Sources for Meeting South Asia's Growing Energy Needs" *Asia Pacific Issues,* No. 83, August.

Singh, J.P. and Pannu, Vinay (2009): "Geopolitics of Energy Cooperation in South Asia", www.visionriviewpoint.com

Chaudhury, Anasua B.R. (2009): "The Energy Crisis and the South Asian Security: An Indian Perspective", *India Quarterly,* 65(2).

Lama, Mahendra P. (1999): "Energy Cooperation in South Asia: Issues, Challenges and Potential", New Delhi: South South Solidarity.

Raina, D.N. (2005): "Regional Energy Cooperation in South Asia: A Vehicle for Poverty Alleviation", Presentation at WEC South Asia Tank Force Forum on Regional Energy Cooperation South Asia, 4th September, Colombo.

Ahmed, Sadiq and Ghani, Ejaz (2007): *South Asia: Growth and Regional Integration*, New Delhi: Macmillan.

Yoshida, Nobou (2008): "A Note on the Trend of Regional Inequality across Areas in the South Asia Region." *Mimeo,* Washington, DC: World Bank.

Research and Information System for Developing Countries (2008): *South Asia Development and Cooperation Report,* New Delhi: Oxford University Press.

International Energy Agency Report, 2009.

16

Emerging Nuclear Order and South Asia

C. Vinodan

The proliferation of nuclear weapon is one of the foremost security challenges facing South Asia in the globalised era. The major concerns in this area are the rise of non-state actors and the fragile state structure in South Asia. Does the crisis in governance and proliferation of nuclear weapons cause nuclear terrorism and nuclear insecurity in the region? The spread of terrorism and transnational criminal networks in South Asia raises this crucial question. Risks of nuclear terrorism and blackmail have increased significantly in recent years mainly because of three factors: the growth and spread of nuclear weapons, the expansion of civilian nuclear programs and the increase in extremist political groups waging campaigns of terror (Saideman and Zahar, 2008).

It is one of the few remaining parts of the world to have failed meaningfully to move its security away from the exercise and manipulation of military threat and towards the building of transparency, cooperation and trust, and the creation of frameworks and institutional arrangements within which security issues can be handled at the political level and without violence. One aspect that has received inadequate attention is the relationship between terrorism and the regional politics of nuclear weapons. The South Asian region is characterized by the active presence of terrorists who have the potential for indiscriminate mass violence, and by the growth of nuclear tensions, particularly after the nuclear tests in 1998 by India and Pakistan (Hersh, 2009).

The longstanding hostility between India and Pakistan has the potential to facilitate nuclear/radiological terrorism in a number of ways. The scope for terrorists to determine the course of events in the region parallels the nuclear stances of India and Pakistan. Each form of nuclear posture carries some risk of terrorist involvement. This growing danger encompasses much of the world. [1] Lax or inadequate security over nuclear materials and weapons in one country could be exploited to trigger atomic blackmail and terrorism elsewhere. Inadequate security at nuclear facilities also could provide extremists waging a campaign of terror within a nation an opportunity to create a situation of national terror by seizing or sabotaging a civilian nuclear power plant or a research reactor or a laboratory.

The increasing level of technological sophistication among terrorist groups, coupled with a renewed determination to achieve political goals, has significantly raised the potential for nuclear terrorism. This paper analyses the growing menace of terrorism and nuclear risks in South Asia. The concerns regarding safety, security and stability of South Asia attracted global attention following the Indo-Pak nuclear stand off and the fragile state structure of Pakistan. The paper argues that link between the terrorist groups and the Pakistani military establishment is posing danger to the region in particular and to the entire world.

Social science research on efforts to maintain safe operations in many modern technological systems suggests that serious accidents are likely over time if the system in question has two structural characteristics: high interactive complexity and tight-coupling. While the Indian and Pakistani nuclear arsenals are small and not complex, it is also clear that the South Asian nuclear relationship is inherently tightly coupled because of geographical proximity. With inadequate warning systems in place and with weapons with short flight times emerging in the region, the time-lines for decision making are highly compressed, and the danger that one accident could lead to another and then lead to a catastrophic accidental war is high and growing.

The proximity of New Delhi and Islamabad to the potential adversary's border poses particular concerns about rapid decapitation attacks on national capitals. Moreover, there are legitimate concerns about social stability, especially in Pakistan, that could compromise nuclear weapons safety and security (Sagan, 2001). Military stability and deterrence in South Asia continue to pose sources of regional and international concern. Despite efforts both within and beyond the region to achieve greater regional stability, conflict between India and Pakistan continues over Kashmir.

In recent years, Islamabad's support for the Taliban has elicited widespread international criticism. And the violent actions which it promotes both within and beyond Pakistan have worked to heighten regional tensions. Beyond this, turmoil in Afghanistan has generated refugee flows, which are raising tensions in Pakistan's border regions. Still, Islamabad's principle source of external concern remains India with past history and geopolitics both playing a central role in its threat calculus (Chakma, 2010). Lacking the territorial depth, economic power and population of its eastern neighbour, and having suffered considerable damage at India's hands in 1971, insecurities run deep—with New Delhi viewed not only as a regional rival but as a deadly threat to the continued integrity of the nation itself.

State Failure and Nuclear Crisis in South Asia

South Asia is passing through an extremely delicate phase in its history. Recent instability in Pakistan, including the Taliban's advance into settled areas, prompted the Pakistani military to undertake large-scale military operations in the Swat Valley. As military and Taliban forces fight in the rugged tribal terrain, several Western analysts have raised concerns about the future of nuclear Pakistan. [2] Now a days the immediate risks involving the proliferation of nuclear weapons is from terrorist threat rather than from state using these weapons in a war front as an ultimate deter. Since the proliferation of nuclear weapons happened across the various regions including South Asia which involved state actors claimed it as foremost for securing its national security.

However, the intentions behind these vary greatly. In such a context, the involvement of rational and irrational actors getting into the sphere of proliferation of WMD looming large. So the risks of nuclear weapons falls into the hands of irrational and non-state actors such as Al-Qaeda remain the principal threat to the safety and stability of South Asia. This is the negative effects of proliferation as against the ideas conceived by proliferation optimists like Kenneth Waltz, [3] who supporting proliferation of nuclear weapons across the regions that decrease the chances or war.

When these nuclear weapons spread across the world the concern of safety and security arises. It needs high degree of command and control and high alerts. In the present situation transnational character of terrorist's organizations can increase the vulnerability of security of nuclear weapons such as theft of fissile material or attack against nuclear facilities. In such a context nuclear deterrence cannot be used against international terrorist's organizations because terrorists do not have land, industries, people, or standing armies that can be targeted retaliatory strikes (Sethi, 2010).

The future goal of the terrorists will push both countries to persuade a nuclear standoff between the two countries which will destabilize the subcontinent. This will led to the favourable environment for non-state actors to indulge in violence. That is why they are continuously organizing terror attacks like the 26/11, 2008 Mumbai terror attack and the 3/3, 2009 terrorist attack on Sri Lankan cricketers in Lahore. The continuing political instability in Pakistan and lack of democratic roots led to the country sides of Pakistan especially North Western Frontier which is sharing its border with Afghanistan fall into the hands of extremists.

This was more evident in the Swat Valley were Pakistan government compromised with Taliban outfits and agreed to implement *Sha-riat-Law* and this was one of the explicit example of the extent of extremist influence in the governance of Pakistan. [4] Islamist extremism and militancy has been a menace to Pakistani society throughout the post-2001 period, becoming especially prevalent since 2007. Pakistan is the site of numerous armed insurgencies of various scales that represent an increasingly severe threat to domestic, regional, and perhaps global security. [5]

With over 10,000 killed in violent incidents across the country in 2010, Pakistan was the most volatile country in the region, pushing war-ravaged Afghanistan—which saw its most lethal year since the beginning of the Global War on Terror in 2001—to second position in this regard. As against 7,123 people killed in Afghanistan through the year, 10,003 were killed in Pakistan in incidents of violence and terrorism. Still, this is an improvement over 2009 when 12,623 fatalities were reported in incidents of violence and terror. [6]

On the other hand, more civilians lost their life in 2010 than the preceding year as terrorist attacks in the Federally Administered Tribal Agencies increased by 28 percent despite ongoing military operations in South Waziristan, Bajaur and Orakzai. Though there has been a significant decline in casualties in incidents of violence in the region stretching from South Asia to West Asia, it is not indicative of any substantial improvement in the regional security situation on the whole. A particularly alarming development in recent years is the significantly increased incidence of militants making direct attacks on Pakistani security institutions. [7]

There was a growing trend of militants garnering support by promising to fill a vacuum left by "ineffective" government structures. [8] The myriad and sometimes disparate Islamist militant groups operating in Pakistan, many of which have displayed mutual animosity in the past, appear to have become more intermingled and mutually supportive in recent times. [9]

The nuclear risks currently present in South Asia, and which are likely to grow and mutate in the coming years, can be broadly categorized as "arsenal risks" and "employment risks." The former stem from the very existence of nuclear arsenals in the two countries and are largely related to the areas of safety and security. The latter are concerned with the employment and seizure of nuclear weapons, either deliberately or inadvertently. [10] In the narrow sense, it connotes the launching of a nuclear strike by one side because of systemic problems in the areas of intelligence, command and control, and decision-making.

The increasing leverage of non-state actors and fragile state structure in Pakistan added much attention to this argument. In the broader sense it also incorporates unplanned but deliberate escalation wherein despite the original intention to keep risk-taking below the nuclear threshold, matters get out of hand because of the escalatory tendency inherent in crisis and war. The "arsenal risks" present in South Asia with regard to safety and security is difficult to evaluate. Both countries put out virtually no information on these matters in the public domain.

More important, it is difficult to estimate how the 1998 shift from a "non-weaponized" posture to a "non-deployed" posture has impacted safety and security issues. The shift has not reduced the opacity of the two arsenals but it is clear that warhead and missile numbers are now rapidly increasing, limited more by production capacity than by international restraints. Pakistan's difficulties in this area are likely to stem from its first-use posture and from its technological and financial constraints. [11]

The "employment risks" in South Asia, as elsewhere, can be broadly categorized as preventive, pre-emptive, inadvertent, and deliberate. Politically, a bolt-from-the-blue preventive strike is very difficult. Operationally, while such a strike might catch many delivery systems in their peacetime concentrations, it would not catch them all. In the case of bombs and warheads, the numbers destroyed are unlikely to be significant. Destroying the bulk of one another's arsenals after the systems have been dispersed on crisis warning is operationally impossible for both countries. [12] Poor intelligence about the opponent's decision to launch a nuclear strike can pose an excruciating dilemma in South Asia.

An inadvertent strike from either side stemming from deficiencies in intelligence, command and control, and decision-making cannot be ruled out. [13] Poor decision-making can also spell danger, particularly in Pakistan where narrow military considerations have tended to dominate strategic decision-making dysfunctionally. [14] To complicate the situation, Pakistan has sought to subvert rational deterrence by trying to sap all its three main components. The rationality of actors is sought to be undermined by projecting a disdain for relative cost calculations, the unitary nature of actors by the possibility of political-military and radical- conservative splits in decision-making circles, and sensitivity to costs by a supposed Islamic nonchalance towards death. The other type of inadvertent war risk—decisions rolling out of control in an escalatory spiral—is also possible in the India-Pakistan context. There is deep hostility, mistrust, poor communications, and, most dangerously, a mutual discounting of one another's resolve and capability.

Nuclear Safety and Stability in Pakistan

Pakistan is a vitally important country in the contemporary global political system. It is a de facto nuclear state, and a pivotal country in the War on Terror. The fragile state structure in Pakistan and the increasing influence of non-state actors in to military and nuclear establishment raises serious concern across the globe. The assassination of Benazir Bhutto on December 27, 2007 was a sharp reminder of the country's institutional collapse. A simple truth in vast regions of Pakistan today is that the state has withered away and the non-state actors are extended their influence in the Swat Valley, North West Frontier Province, federally administered tribal areas and Baluchistan.

The security of Pakistan's nuclear weapons and infrastructure has been the subject of much coverage and debate in recent months as Pakistani government forces have stepped up their fight against insurgents. Growing extremism, an expanding nuclear portfolio, and continuing instability challenge Pakistan's ability to protect its nuclear arsenal. Non-proliferation experts and researchers warns in particular against the slow leak of nuclear expertise and materials to Non State actors. [15] The state of Pakistan also failed to address the new challenge in a serious manner. The other stake holders India and the United States failed to address the issue in a meaningful way.

A potentially much greater threat comes from flawed assumptions and rhetorical excesses, which could lead Pakistan, India and the United States down the wrong path. Pakistan's nuclear program is nearly three decade old (Albright, 2000). Guided by its animosity towards and rivalry with India, Pakistan's nuclear weapons capability is aimed at matching and undermining India's regional influence. Domestic imperatives play a significant role in shaping Pakistan's nuclear policy. Pakistan's nuclear weapons program is the preserve of its anti-Indian military establishment. Its political bureaucracy assumes a subordinate role. Even during Pakistan's brief democratic interludes, the military retained control over the nuclear weapons programme.

The country became an overt nuclear weapon power in 1998, when it carried out several tests in response to India's resumption of nuclear tests. In recent years, Pakistan has continued to expand its nuclear weapons program both in terms of delivery systems and through more effective warheads by proceeding from highly enriched uranium (HEU)-based devices to those based on plutonium. [16] According to analyses in 2010, a second plutonium production reactor is likely to have begun operations by late 2009. [17]

Construction of a third plutonium production reactor at Khushab began in 2006. In addition, associated facilities are being expanded, especially plutonium separation facilities at New Labs, Pakistan Institute of Science and Technology, to reprocess spent fuel from the new reactors at Khushab (Albright and Brannan, 2009). Thus, Islamabad appears to be expanding and diversifying its nuclear weapons capability through the plutonium route's more compact and powerful warheads in its quest for a secure second-strike capability against India (Nelson, 2006). There is no confirmed data regarding the location and production sites of Pakistan's nuclear weapons, but Islamabad is considered to have more than half-a-dozen storage locations. [18]

It is commonly believed that the fissile cores of Pakistan's arsenal are stored separately from the non-nuclear components. [19] However, during the Kargil conflict, Pakistan is believed to have armed its nuclear capable missiles in possible preparation for nuclear strike on India, and Islamabad has hinted at having considered the nuclear option during the Indo-Pak confrontation in 2002. [20] None of the Pakistani nuclear materials is under International Atomic Energy Agency Safeguards. The enrichment of weapon grade material has progressed and the number of Pakistan's nuclear weapons has increased since the initial weapons tests in 1998.

In 2002, after a year of frequent tensions with India, Pakistan has reported to have accelerated work on its nuclear arsenal, with workers putting in around-the-clock shifts. [21] Estimated Pakistan's nuclear strength before 9/11 was about 425-800 kilogram (kg.) of weapons-grade uranium, about 600 kg. of plutonium and up to 7 dozens nuclear weapons. [22] A recent report suggests that Pakistan possess up to 70-90 nuclear weapons and has fissile material for up to 52 more. Thus, one can assume that Pakistan's nuclear arsenal will continue expand at a rapid pace, and the requirements for securing it will concomitantly increase.

The concern regarding the safe custody of Pakistan's nuclear weapons have arisen because of the steady growth of Jihadi culture supported by the ruling establishment. The process started during the regime of Zia-ul-Haq who promoted Nizam-e-Mustafa in the country; American sponsored against the Soviet forces in Afghanistan accelerated the spread of religious extremism in Pakistan. Brigadier S.K. Malik *The Quaranic Concept of War*, with forward by General Zia-ul-Haq, is a notorious example of religious concept being imported in to the theory and practice of nuclear deterrence (Zuberi, 2001). It is generally believed that the military, as the important player in the nation, has full responsibility for command and control of Pakistan's nuclear weapons.

However, after the events of September11, 2001, there was in international circles considerable concern about, and analysis of, the safety of the Pakistani nuclear arsenal. [23] There were reports that nuclear arsenal had been dispersed to multiple locations after the September 11 attacks. [24] Dispersion could itself result in an increased internal threat to the arsenal by increasing the number of locations that could be targeted by terrorists. The Musharaff government was quick to dismiss concerns about the Pakistani arsenal. Musharaff himself claimed that it was safe and that there was no chance of its falling in to the hands of fundamentalists.

However, given Pakistan's history and its extensive and persistent links to Islamic radicals, such claims need to be viewed with some scepticism. Potential dangers to Pakistan's nuclear assets include loss of control over nuclear technology, weapons, weapons components or fissile material. The most commonly discussed threat is from terrorists or their sympathizers within Pakistan military and nuclear establishments. In addition, a coup could instantly transfer control of Pakistan's nuclear assets to a new regime with unknown intentions. There is also the possibility (albeit a relatively remote one) of the Pakistani establishments knowingly passing some of its nuclear assets to another nation, or to a non-state actor. Under any of those circumstances, the dangers to region security would increase. Possession of nuclear weapons by terrorists is an especially grave concern in

India after the December 13, 2001, terrorist attack on the Indian Parliament.

A.Q. Khan Network and Nuclear Terrorism

After the events of September 11, 2001, no one doubts that terrorists might be interested in killing a lot of people. In our effort to understand the level of risk of nuclear terrorism in the future, it is therefore worth considering the relationship, if any, between the spread of nuclear weapons to fragile states and the danger that terrorist organizations will get and use nuclear weapons. Some terrorists, like Osama bin Laden and the Al-Qaeda network, have been quite open in stating their desire for nuclear weapons.

Indeed, after Osama bin Laden declared a Jihad (holy war) against the United States, he was asked about reports that he wanted nuclear weapons and replied, "...to possess the weapons that could counter those of the infidels is a religious duty." Any terrorist leader with this kind of strategic vision is not likely be deterred from using nuclear weapons or radiological weapons against the moderate states. Pakistan is clearly the most serious concern in this regard. Prior to September 11, there were no specialized Pakistani teams trained on how to seize or dismantle a stolen nuclear weapon.

No dedicated personnel reliability program (PRP) was in place to insure the psychological stability and reliability of the officers and guards of Pakistan's nuclear forces. The danger of terrorists gaining access to a Pakistani nuclear weapon is heightened during crises when Pakistan is likely to go on a nuclear alert and disperse weapons from secure storage sites to make them invulnerable to an Indian attack.

Dispersal, however, makes such weapons more vulnerable to attack or seizure by terrorists or terrorists aided by military insiders. The existence of this "vulnerability/invulnerability paradox" should provide a strong incentive to avoid military threats and crises in South Asia (Sagan, 2001). In February 2004, Dr. Abdul Qadeer Khan, the former head of Khan

Research Labs, confessed to the illicit transfer of nuclear weapons technology to Iran, Libya, North Korea and other countries. Khan is reported to have amassed millions of dollars from the transfers.

According to media reports, Khan reportedly told Inter Services Intelligence officials that he transferred nuclear weapons technology so that other Muslim countries could use it to enhance their security. [25] Khan was pardoned by President Musharraf in the wake of these revelations for his involvement in smuggling, in return for cooperating with Pakistani authorities to uncover the extent and scope of the smuggling network. Despite his pardon, Khan remains under house arrest while subject to questioning by Pakistani security services. In addition, several scientists and security officials have been detained and questioned, but the Pakistan government did not take any concrete action against any of these groups and individual including Khan. [26]

It was reported that Khan used his national status as one of the leaders of Pakistan's nuclear efforts to conduct his black market sales of nuclear material and technology. He had established black market channels initially to acquire nuclear technology for Pakistan, but slowly the direction of the flow reversed. His old black market channels became the new channels by which Khan distributed nuclear technology and material to Iran, North Korea, and Libya. Khan did not appear to take advantage of the loopholes in the old Pakistani export control system. He had little or no oversight from the government.

According to an aide close to President Musharraf, "Khan had a complete blank check. He could do anything. He could go anywhere. He could buy anything at any price." [27] In the aftermath of these revelations concerning the black market smuggling network of A.Q. Khan, the Pakistani Ministry of Foreign Affairs prepared new export control regulations to address future concerns raised by the International Atomic Energy Agency and the United States. [28]

On June 7, 2004, the Pakistani government introduced a bill in the National Assembly, known as the Export Control on Goods, Technologies, Material and Equipment related to Nuclear and Biological Weapons and their Delivery Systems Act, 2004. The bill was subsequently passed by the Pakistan National Assembly and Senate on September 14 and 19, respectively. [29] The bill stipulated that any violation of the act would result in up to 14 years' imprisonment, forfeiture of all property and assets to the government, and a fine of 5 million rupees (about US$ 86,500).

Any individuals attempting to commit or abetting the commission of such offences would be charged as if they had themselves committed the violation. [30] The act also called for the creation of an oversight board to administer export control regulations, enforcement of the act, and licensing for export and re-export of nuclear-and biological-related goods and technology. Furthermore, exporters will also be required to maintain records of all transactions and report them to the designated government agencies. All government agencies involved in the licensing process will be required to maintain records of all recommendations and decisions involving licensing. The control lists of items subject to licensing requirements will be reviewed periodically and updated as required by the government. [31]

In spite of all these legislations and rhetoric by the Pakistani military and civilian leadership, the international community remains concerned with the growing menace of nuclear terrorism in South Asia. In the late 2007 and early 2008, some observers expressed concern about the security of Pakistan's nuclear arsenal if political instability were to persist. [32] Former Prime Minister late Benazir Bhutto said in a November 5, 2007, interview that, although then President Musharraf claimed to be in firm control of the nuclear arsenal, she feared this control could weaken due to instability in the country. [33]

Similarly, Michael Krepon of the Henry L. Stimson Centre

has argued that “a prolonged period of turbulence and infighting among the country’s President, Prime Minister, and Army Chief” could jeopardize the army’s unity of command, which “is essential for nuclear security.” [34] During that time, US military officials also expressed concern about the security of Pakistan’s nuclear weapons. [35] Director General of the International Atomic Energy Agency (IAEA), Mohamed El-Baradei, also has expressed fears that a radical regime could take power in Pakistan, and thereby acquire nuclear weapons. [36] Experts also worry that while nuclear weapons are currently under firm control, with warheads disassembled, technology could be sold off by insiders during a worsened crisis. [37]

However, US intelligence officials have expressed greater confidence regarding the security of Islamabad’s nuclear weapons. Deputy Secretary of State John D. Negroponte in testimony to Congress on November 7, 2007 said he believed that there is “plenty of succession planning that’s going on in the Pakistani military” and that Pakistan’s nuclear weapons are under “effective technical control.” [38] Similarly, Donald Kerr, Principal Deputy Director of National Intelligence, told a Washington audience May 29, 2008, that the Pakistani military’s control of the nuclear weapons is “a good thing because that’s an institution in Pakistan that has, in fact, withstood many of the political changes over the years.”

A Department of Defence spokesperson told reporters December 9, 2008, that Washington has “no reason at this point to have any concern with regards to the security” of Islamabad’s nuclear arsenal. More recently, Maples stated March 10, 2009, that Islamabad “has taken important steps to safeguard its nuclear weapons,” although he pointed out that “vulnerabilities exist.” Secretary of Defence Robert Gates stated in a January 21, 2010, interview that the United States is “very comfortable with the security of Pakistan’s nuclear weapons.” [39]

Other governments have also voiced opinions regarding

the security of Pakistan's nuclear arsenal. For example, former Indian National Security Adviser M.K. Narayanan said that the arsenal is safe and has adequate checks and balances. [40] Similarly, Secretary of State for Foreign and Commonwealth Affairs David Miliband told the *Charlie Rose Show* December 15, 2008 that Islamabad's nuclear weapons "are under pretty close lock and key." Russian Deputy Prime Minister Sergei Ivanov, however, sounded somewhat less optimistic in a March 24, 2009, television interview, stating that Moscow is "very much concerned" about the security of Pakistan's arsenal. [41] Pakistani officials have consistently expressed confidence in the security of the country's nuclear arsenal.

Then President Musharraf stated in November 2007 that Pakistan's nuclear weapons are under "total custodial controls." [42] President Asif Ali Zardari told CNN on December 2, 2008, that the country's nuclear command and control system "is working well." Additionally, a Pakistani Foreign Ministry spokesperson stated on May 21, 2009, that "there is simply no question of our strategic assets falling into the wrong hands. We have full confidence in our procedures, mechanisms and command and control systems."

In addition to the above scenarios, the security of Pakistan's nuclear weapons could also be jeopardized by another conflict between India and Pakistan, Michael Krepon argued, explaining that an "escalating war with nuclear forces in the field would increase the probability of accidents, miscalculations, and the use of nuclear weapons." This is because when tensions rise precipitously with India, the readiness level of Pakistan's nuclear deterrent also rises.

Because the geographical coordinates of Pakistan's main nuclear weapon storage sites, missile, and air bases can be readily identified from satellites—and therefore targeted by opposing forces—the dictates of deterrence mandate some movement of launchers and weapons from fixed locations during crises. Nuclear weapons on the move are inherently less secure than nuclear weapons at heavily-guarded storage sites.

Weapons and launchers in motion are also more susceptible to "insider" threats and accidents. [43] Such a war, Krepon added, would also place stress on the army's unity of command. Krepon has also pointed out that Islamabad faces a dilemma, because less-dispersed nuclear weapons may be more vulnerable to a disarming military strike from India. [44]

Conclusions

Although nuclear weapons may no longer be at the forefront of great power diplomacy with the familiar Cold War confrontation between the United States and the Soviet Union fading into history, nuclear weapons remain an important consideration for international policy makers. Even though Cold War-era deterrence strategies fall short when faced with non-state nuclear threats because of the underlying state-centric assumptions, the lack of rational motives exhibited by certain non-state groups and the problems of retaliating against an actor with no territorial base, the theoretical framework on which Cold War deterrence rests still offers much advice to policy makers today even when considering threats from non-state actors.

Seeking to deter terrorists, especially committed, jihadi groups such as Al-Qaeda willing to use weapons of mass destruction (WMD), poses significant challenges. In this context, India and Pakistan must realise their strategic vulnerabilities in the emerging global security scenario. India should play a meaningful role in redefining the security structure in the subcontinent. It should accept the challenges facing Pakistan for its survival in the post-Iraq war scenario. A stable, democratic and prosperous Pakistan will certainly enhance stability and security of the subcontinent. Any attempt to destabilize Pakistan either by India or by external powers would certainly destabilize the region or jeopardize the security interests of India. Pakistan should also recognise and accept its strategic vulnerability in the emerging international order.

Pakistan's immediate concerns will be to ensure the safety and security of its nuclear establishment along with containing terrorism in their soil. Any further attempt to collaborate with the Jihadi groups or proliferation efforts would certainly invites external intervention in to the country. A mature democratic leadership is needed to tackle the most volatile situation in the history of Pakistan and also in the history of the subcontinent. Prime Minister Manmohan Singh and the Pakistani leaders (Zardari/Gilani) now have a rare opportunity to engineer the shift from recurring crises to a nuclear safety. The immediate task is to ensure nuclear safety, stability and security in South Asia. This can be achieved through improved bi-lateral relations and the avoidance of another crisis.

India and Pakistan should sincerely review the existing nuclear confidence and security building measures and stabilize the existing CBMs. And without nuclear CBMs the dangerous political tensions between the two countries, fuelled by feelings of mutual vulnerability, cannot be mitigated. The time may have come to develop a new balance between the needs of war risk reduction and those of non-proliferation. It is the responsibility of the Indian and Pakistani leadership to come out of their negative nationalistic perspective to a broad national, region and global approach to fulfil the long cherished aspirations of the millions of the people in the sub-continent for peace, prosperity and development.

End Notes

1. See Leonard Spector, "Pakistan, Taliban and Global Security-Part I", *Yale Global*, May 8, 2009; Riedel, "Pakistan and the Bomb"; David Albright, Paul Brannan, and Robert Kelley, "Pakistan Expanding Dera Ghazi Khan Nuclear Site: Time for US to Call for Limits", *ISIS Imagery Brief*, May 19, 2009, p. 4.
2. See Bruce Riedel, "Pakistan and the Bomb", *The Wall Street Journal*, May 30, 2009; Bryan Bender, "Pakistan, US in Talks on Nuclear Security", *Boston Globe*, May 5, 2009, www.boston.com/news/nation/washington/articles/2009/05/05/pakistan_us_in_talks_on_nuclear_security/; Andrea Shalal-Esa, "U.S. Lawmaker Worried by Sag in Support for Defense",

Reuters, June 24, 2009 (quoting Representative John Murtha [D-Pa.] about US air strikes against Pakistani nuclear targets.

3. As a body of thought, nuclear optimism has passed through two distinct stages. Although its lineage can be traced to the classic deterrence theorists, its first robust articulation was Kenneth Waltz's iconoclastic "more may be better" argument. Waltz uses rational deterrence theory and structural realism to advance two interconnected propositions. First, "nuclear weapons, responsibly used, make wars hard to start. Nations that have nuclear weapons have strong incentives to use them responsibly", (Kenneth N. Waltz (1981), The Spread of Nuclear Weapons: More May Be Better", Adelphi Papers 171 (International Institute for Strategic Studies, 30). Second, the first proposition holds true "for small as for big nuclear powers;" thus, "the measured spread of nuclear weapons is more to be welcomed than feared" (Waltz, "The Spread of Nuclear Wepons,"; 30)
4. See for more discussion on the related issues Brahma Chellaney "Fighting Terrorism in Southern Asia: The Lessons of History", *International Security,* 23 (3) winter 2001/02, pp. 94-116. Thomas Fuller, "Stroking Southeast Asia Tensions," *International Herald Tribune*, October 31, 2001, p. 1.
5. For example, in late 2008, the British Prime Minister estimated that three-quarters of the most serious terrorism plots investigated in Britain had links to Al-Qaeda in Pakistan; "Brown Offers Pakistan Anti-Terror Aid," *Washington Post*, December 15, 2008.
6. Pakistan Security Report (2010), PIPS, Islamabad. The report shows that there was also a decline in casualties among "militants" despite a 165 percent increase in drone attacks that are supposed to specifically target terrorists based on intelligence. "Militants' casualties" fell by 27 percent since 2009.
7. In recent examples, a March suicide car bombing of the facilities of a special counterterrorism investigate unit in Lahore killed at least 15 people and destroyed the entire building. Days later, twin suicide attacks on other army targets in the same city killed dozens more ("Suicide Car Bombers Strikes Pakistani Intelligence Unit," *Los Angeles Times*, March 8, 2010; "Twin Suicide Attacks Hit Military Sites in Pakistan," *New York Times*, March 12, 2010).

8. Brian Fishman, "The Battle for Pakistan: Militancy and Conflict Across the FATA and NWFP," New America Foundation, April 2010.
9. "Pakistan Attacks Show Militants' Reach," *Chicago Tribune*, October 16, 2009; Asif Ali Zardari, "Democracy is the Greatest Revenge", *Wall Street Journal*, December 27, 2009.
10. For the inadvertent nuclear risks facing small nuclear powers, see Hakan Wiberg, Ib Damgaard Petersen, and Paul Smoker, (eds.), *Inadvertent Nuclear War: Implications of the Changing Global Order,* Oxford: Pergamon, 1993.
11. For the dangers present in this area, although largely in the Cold War setting, see Scott D. Sagan, *The Limits of Safety: Organizations, Accidents and Nuclear Weapons* (Princeton, N.J.: Princeton University Press, 1993) and Bruce G. Blair, *The Logic of Accidental War,* Washington D.C.: Brookings, 1993.
12. "India's and Pakistan's invulnerable nuclear capabilities make it unlikely that either would plan a pre-emptive strike," Neil Joeck, *Maintaining Nuclear Stability in South Asia*, p. 56. It is also noteworthy that not a single mobile Scud launcher could be destroyed during the Gulf War, according to Mark Hiller of Johns Hopkins University quoted in Stewart M. Powell, "Scud War, Round Three," *Air Force Magazine*, October 1992, p. 33.
13. For inadvertent war risks see Barry P. Posen, *Inadvertent Escalation: Conventional War and Nuclear Risks* (Ithaca, N.Y.: Cornell University Press, 1999); Robert Jervis, *The Meaning of Nuclear Revolution: Statecraft and the Prospect of Nuclear Armageddon,* Ithaca, N.Y.: Cornell University Press, 1989; and Richard Smoke, *Controlling Escalation, Cambridge,* MA: Harvard University Press, 1977.
14. See Timothy D. Hoyt, "Pakistan's Nuclear Doctrine and the Dangers of Strategic Myopia," *Asian Survey* XLI: 6, November/December 2001, pp. 1064-1086; Ahmed Faruqui, "Pakistan's Strategic Myopia," *RUSI Journal*, April 2000; Altaf Gauhar, "Four Wars, One Assumption"; and Julian Schofield, "Militarized Decision-making for War in Pakistan: 1947-71.".
15. ACA July 2, 2009.
16. Robert S. Norris & Hans Kristensen, "Nuclear Notebook: Pakistani Nuclear Forces, 2009," *Bulletin of the Atomic Scientists*, September/October 2009, http://thebulletin.metapress.com.

17. In 1998, Pakistan had commissioned its first plutonium production reactor, also at Khushab, which is capable of yielding 10-15 kg. of weapons-grade plutonium annually. Paul Brannan, "Steam Emitting from Second Khushab Reactor Cooling Towers; Pakistan Maybe Operating Second Reactor", Institute for Science and International Security, March 24, 2010, http://isis-online.org. Also see Zia Mian, "Pakistan May Have Completed New Plutonium Production Reactor, Khushab-II" International Panel on Fissile Materials, February 28, 2010, www.fissilematerials.org.
18. Robert S. Norris and Hans M. Kristensen, "Nuclear Notebook: Worldwide Deployments of Nuclear Weapons, 2009", *Bulletin of the Atomic Scientists*, November/December 2009, http://thebulletin.metapress.com.
19. David Albright, Kevin O'Neill and Corey Hinderstein, securing Pakistan's Nuclear Arsenal: Principles for Assistance, Institute For Science and International Security ISIS Issue Brief, October 4, 2001, 'Pakistan Prepared Nuclear Strike', *BBC News Online*, May 16, 2002; and 'Mushraff Hints He Considered Nuclear Strike', *Washington Post*, 13 December, 2002, p. 11.
20. Roy McCarthy, 'Rivals Compete in Daily race to set up Nuclear Arms Production', *Guardian*, May 27, 2002.
21. David Albright, "Securing Pakistan's Nuclear Weapons Complex", *Paper delivered at Stanely Foundation Peace Conference*, October 25-27, 2001. Also see David Albright, Kevin O'Neill and Corey Hinderstein, securing Pakistan's Nuclear Arsenal: Principles for Assistance, Institute for Science and International Security ISIS Issue Brief, October 4, 2001.
22. For example, see P. Cotta Ramusino and M. Martellini, *Nuclear Safety, Nuclear Stability and Nuclear Strategy in Pakistan*, Ladan Network, Centro Volta Corno, Italy, February 2002, www.nautilus.org/pakistan.doc.
23. 'Pakistan's Nuclear Weapons were moved', *Global Security Newswire*, 14 November 2001; also see Molly Moore and Kamran Khan, 'Pakistan Moves Nuclear Weapons; Musharraf says Arsenal Now Safe', *Washington Post*, 11th November, 2001.
24. "A.Q. Khan confesses to leaking nuclear secrets to Iran, Libya," Agence France-Presse, February 2, 2004.
25. "Powell Says Nuclear Ring Broken," BBC News, February 9,

2004, http://www.bbc.co.uk/news; "Pakistan Scientists to Face Nuclear Charges Selling Nuclear Secrets," The Independent, January 27, 2004, on Lexis-Nexis.

26. William J. Broad, David E. Sanger, and Raymond Bonner, "How Pakistani Built His Network," New York Times, February 12, 2004, http://www.nytimes.com.
27. "Pakistan: Upcoming Nuclear Bill May Empower Govt to Punish Proliferators," The News (Islamabad), June 5, 2004.
28. "Pakistan Assembly Passes Bill to Control Nuclear Goods Export," Islamabad APP, September 14, 2004, in FBIS IAP20040914000068; "Pakistan Senate Adopts Bill on Nuclear Export Controls," Islamabad PTV World, September 18, 2004.
29. Shahid Hussain, "Islamabad Tightens Nuclear Controls," Gulf News, May 6, 2004, on Lexis-Nexis, http://www.lexis-nexis.com; "Govt Introduces Bill to Control Proliferation of Nuclear Materials," The News (Islamabad), June 8, 2004.
30. "Nuclear Exports Control Bill Tabled in NA," The Pakistan Newswire, June 7, 2004, on Lexis-Nexis, http://www.lexis-nexis.com.
31. "Opinions Mixed on Pakistani Nuclear Security," *Global Security Newswire*, November 6, 2007.
32. Also see comments by David Albright in the same interview ("Pakistan in Crisis: Interview with Benazir Bhutto," *CNN*, November 5, 2007).
33. "US-Pakistan Strategic Relations," Statement before the Committee on Senate Homeland Security and Governmental Affairs Sub-committee on Federal Financial Management, Government Information, Federal Services, and International Security June 12, 2008.
34. "Lieutenant General Carter Ham Holds a Defense Department Briefing," *CQ Transcripts*, November 7, 2007.
35. "Al-Baradei to Al-Hayat," *Dar Al Hayat*, English Edition on-line, January 10, 2008, http://english.daralhayat.com/Spec/01-2008/Article-20080110-639032eb-c0a8-10ed-01ae-81ab2ea588db/story.html.
36. Also see comments by David Albright in "Pakistan in Crisis," 2007.
37. House Foreign Affairs Committee Hearing on Democracy, Authoritarianism and Terrorism in Contemporary Pakistan, November 7, 2007.

38. Secretary of Defense Robert Gates, "Express TV Interview," January 21, 2010.
39. "Pak Nukes Safely Guarded, Says Narayanan," *The Press Trust of India*, December 16, 2007.
40. Lyubov Pronina and Ellen Pinchuk, "Russia 'Concerned' About Security of Pakistan's Nuclear Arsenal,"*Bloomberg*, March 25, 2009.
41. "Pakistan Nukes Under Control: Musharraf," *Agence France Presse*, November 13, 2007.
42. Krepon, June 12, 2008.
43. Michael Krepon, "Complexities of Nuclear Risk Reduction in South Asia," *The Hindu*, May 29, 2009.

References

Sagan, Scott D. (2001): "The Perils of Proliferation in South Asia" *Asian Survey*, November/December, Vol. 41, No. 6.

Hersh, Seymour M. (2009): "Defending the Arsenal: In an Unstable Pakistan, Can Nuclear Warheads be Kept Safe?" *The New Yorker*, November 16.

Chakma, Bhumitra (2010): *South Asia's Nuclear Security,* Routledge.

Sethi, Manpreet (2010): *Nuclear Strategy—India's March Towards Credible Deterrence*: KW Publishers.

Albright, David (2000): "Supplement on Fissile Material and Nuclear Weapons in India and Pakistan", in Joseph Cirincione (ed.), *Repairing the Regime*, New York, 147-149.

Albright, David and Paul Brannan (2009): "Pakistan Expanding Plutonium Separation Facility near Rawalpindi," Institute for Science and International Security, May 19, www.isis-online.org.

Nelson, Dean (2006): "Pakistan Upgrades Nuclear Arsenal", *The Times,* July 30, www.timesonline.co.uk.

Zuberi, Martin (2001): "Safety of Pak Nukes", *World Focus*, New Delhi, October-December.

Saideman, Stephen M. and Marie-Joelle J. Zahar (2008): *Intra-State Conflict, Governments and Security: Dilemmas of Deterrence and Assurance*, New York: Routledge.

Index

Index